This Bud's for You

Legal Marijuana: Selecting, Growing & Enjoying Cannabis

Ed Rosenthal

This Bud's for You

Legal Marijuana: Selecting, Growing & Enjoying Cannabis

Ed Rosenthal

Quick American Publishing

This Bud's For You — Legal Marijuana: Selecting, Growing and Enjoying Cannabis
Copyright © 2017 Ed Rosenthal
Published by Quick American Publishing
A Division of Quick Trading Co.
Piedmont, CA, USA

Printed in Canada
FIRST PRINTING

ISBN: 978-1-936807-30-7
eISBN: 978-1-936807-31-4
Executive Editor: Ed Rosenthal
Project Director: Jane Klein
Project Manager: Darcy Thompson
Chief Editor: Ellen Holland
Editorial Assistance: Sidney Borghino
Photo Editor: Darcy Thompson, Christian Petke
Art Direction: Christian Petke
Design: Scott Idleman/Blink
Additional Design: Stephen Lam
Photographs of varieties appear courtesy of the seed companies. All other photos by Ed Rosenthal.
Front cover photograph courtesy of Greenthumb Seeds of Canada. Back cover photo courtesy of
Nadim Sabella.

Library of Congress Control Number 2016958769

Acknowledgements

Timothy Anderson, Rolph Blythe, Sidney Borghino, Randi Boyce,
Cannabis Now Magazine, Frenchy Cannoli , The Dank Duchess,
Susan Cohen, Jimi Devine, William Dolphin, David Downs,
Dynasty Genetics, Eugenio Garcia, Mark Gray, Ellen Holland,
Scott Idleman, Jane Klein, Gracie Malley, Justin McIvor, Shelli Newhart,
Christian Petke, Harry Resin, Nadim Sabella, Jason Schulz,
Steep Hill Labs, Nikka T, Darcy Thompson, Women Grow, Greg Zeman

"What shall we say, shall we call it by a name
As well to count the angels dancing on a pin
Water bright as the sky from which it came
And the name is on the earth that takes it in
We will not speak but stand inside the rain
And listen to the thunder shout
I am, I am, I am, I am"

Lyrics from the song Weather Report Suite: Part 2 (Let It Grow)
Written by John Perry Barlow; Music by Jerry Garcia
Reproduced by arrangement with Ice Nine Publishing

The Industry Today

[photo by Justin McIvor]

The Breeders and Their Strains

Indica Dominant (70%>)

Hybrid (Indica and Sativa)

[photo by Nadim Sabella]

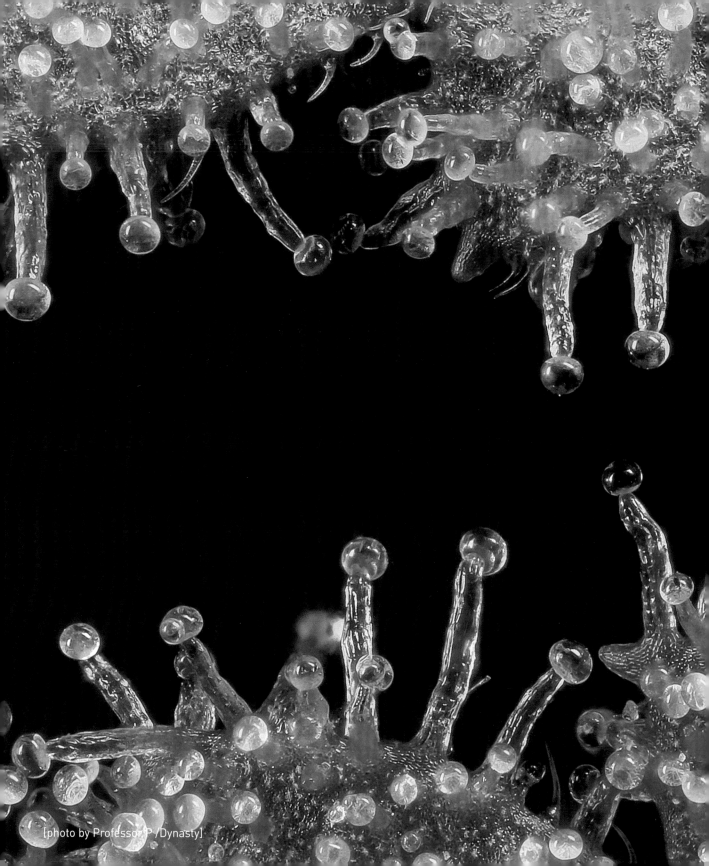

[photo by Professor P /Dynasty]

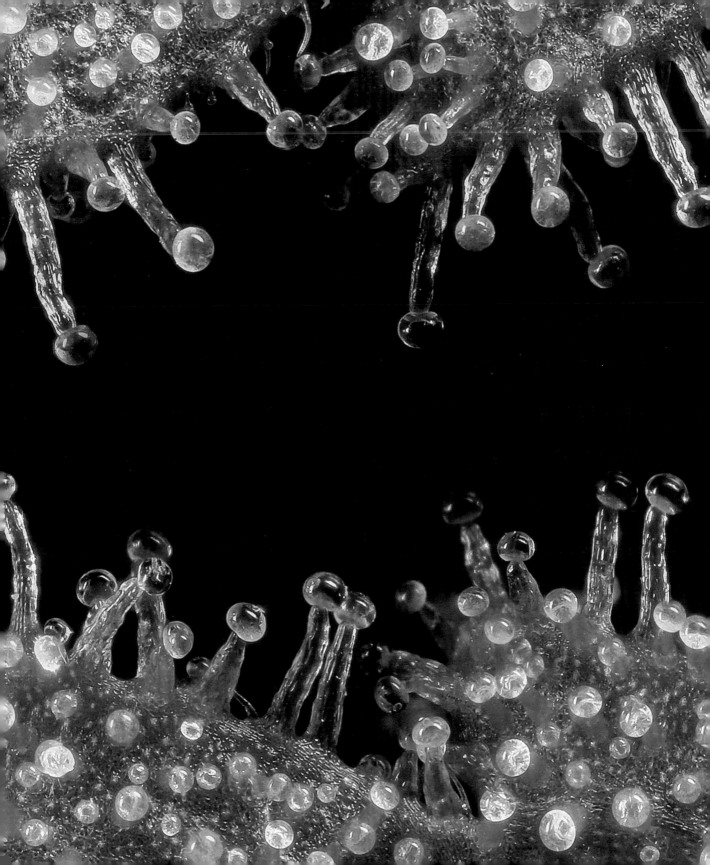

Introduction by Ed Rosenthal

It has taken an exhaustive struggle of more than four decades, more time than most of you have been alive, to achieve even a modicum of Amorphia's original goal of free legal backyard marijuana. This was proposed in California's Prop 19 (1972).

Still, this book is a celebration. We are living in an historical moment. The change from marijuana as pariah, to enthusiasm for its medical potential and the acceptance of its recreational use.

It's a dramatic shift. In one of the few programs aired on the subject in the 1970's the host and panelists all agreed that legalization was not open to discussion—just control and repression. A decade later legalization was still not taken seriously. When I was invited to a popular TV talk show, Morton Downey Jr., the host, sat on my lap trying to stop me from demanding legalization. That didn't work. Nor did the fake news, propaganda, scare tactics, jailings, destruction of reputations and property. Marijuana Activists helped, but Marijuana Won The War. There is a simple reason: Unlike politicians, good marijuana tells no lies. Marijuana also wins more votes.

This Bud's For You is divided into two sections. One portion features interesting articles about the contemporary marijuana environment and some of the people involved in it. They reflect the change in the marijuana culture as it develops into an industry vying to become part of mainstream society. Together, they represent the broad pattern of transformation by people who are not just observers, but part of the commercial surge.

This change happened without laws and regulations being imposed by governments. In 1996 when Prop. 215, the first medical marijuana initiative, was approved by the voters of California, the initiative included nothing about sales and distribution. Rather it was work of pioneering activists and the free market that resulted in quality products—flowers labeled with stats on potency and purity, edibles of consistent doses, and products high in CBD.

The second section is a survey of new varieties that are commercially available. The choice of products starts with the varieties that have expanded beyond Indicas and Sativas to include growing interest in varieties with high levels of CBD. The selections here are part of the new wave. Most of the breeders represent new companies. Some are using traditional artisanal techniques while others are introducing a more scientific approach. Many of the established companies have turned to a new generation of breeders to continue their lines.

These approaches all contribute to the large number of choices the home grower, market gardeners and farmers enjoy. The result is a broader choice for everyone who uses marijuana medically and recreationally.

What you will find in this book is what legal marijuana looks like. It includes selection, sophistication, transparency. Seed money sprouts into a booming industry that encompasses technology, craftsmanship, and connoisseurs. Enjoy the difference.

[photo by by Dabsel Adams]

Durban Poison [photo by Gracie Malley]

SMART Breeding: Growing the Next Elite Strain

Legalization and new techniques in cannabinoid science usher in a new age of marijuana.

By Timothy Anderson

Being a buyer at Harborside Health Center, a bustling medical cannabis dispensary in Oakland California, offers me the unique and unprecedented opportunity to view a diverse spectrum of cannabis of all varieties grown with countless different indoor and sungrown methodologies. But with more choices available than ever before, growing a strain that appeals to a wide range of cannabis consumers occurs only with the melding of art and science.

In an era where it has become easier both legally and technologically for gardeners to cultivate cannabis, the amount of product available in markets like ours is immense but repetitious.

Having access to excellent genetics is as easy as strolling into the local dispensary to choose a strain either from racks overflowing with verdant clones or seed catalogs that are large as the ones in the Netherlands. Although their choices are myriad, growers gravitate toward the most well-known or popular strains and mostly ignore the rest. The result is that both large commercial cultivators striving for dependable and bulletproof plants and newbie growers who only grew a few clones are likely to have similar strains.

Access to desirable genetics for both high-level cultivators and novice growers has become more and more equal. For example, the variety Gorilla Glue #4 went from being exclusive and highly desired to market saturation in a period of less than two years. Compare that to the previous crazes like Girl Scout Cookies, which took several years to reach a wide availability for the Chem/OG/Sour family, which remained tightly controlled for a decade becoming widely cultivated. The days of $1,000 OG Kush clones

Master Kush [photo by Nadim Sabella]

are not yet a distant memory. It is fairly easy to correlate this access with both the legal tolerance of dispensaries as well as the cannabis community's vigorous networking on online forums, and we can only expect the rate of cycling from "rare" to "available" to increase rather than decrease.

Cultivators frequently ask me what they should grow to guarantee a niche on our shelves, and my answer frequently disappoints: "Something we don't see all the time." It should be no surprise that growing from seed is the best way to do this, allowing the grower to pick a phenotype that is both unique and all their own, at least until the grower gives someone else the cutting. Wide retail access to reliable clones in legal states has done much to dampen cultivators' interest in sprouting seeds — let alone saving a male to make their own hybrids — and that has much to do with this slump of diversity in the current marketplace as well.

But the monotonies are easing up a bit. At Harborside the Chems, Cookies and purple Indicas remain the best sellers but hybrids of older African landrace varieties such as Red Congolese and Nigerian Silk have gained popularity and have both modern THC potency levels as well as unique terpene sets. The return to older gene pools is where the search for newness often takes us.

It would not be surprising to see future combinations of modern pre-potent varieties and older, unexplored landraces as breeders search for the promise of new offerings.

The future promises many surprises for cannabis users. The most interesting development: the hyper evolution of cannabis concentrates. This brought about a "second coming" of hashish culture albeit in a new and modern form. Cannabis concentrates are more popular than ever, and the diversity of type continues to multiply. The potency and purity of these concentrates is greater than previously available because extractors have better access to both high-quality input product and cutting-edge technology.

[photo by Elemental Wellness Center]

A new development phenomena is seed breeders and concentrate makers working symbiotically. The resin-covered shattered bracts left over from seed collection are perfect for extractions and collected when green. Some seed producers have created brands for their own extracts to take advantage of their labor. What is most interesting about seed breeders extracting is the potential influence of extract making on breeding. Middle Eastern farmers once bred for traits such as gland size or for the gland to break off from its stalk, this is essential for plants intended for hashish production. Modern breeders growing specifically for concentrate production have the same thoughts, as whole crops of cannabis plants are extracted. These breeders may also tinker with traits such as toughness of the resin gland's cuticle or thickness of its wall. This would be a major advantage to extract makers.

Terpenes are another component of the resin that is being actively explored and bred for. Ed Rosenthal's seminal article in *Big Book of Buds vol. 3* published in 1998 followed by Dr. Ethan Russo's 2011 paper on "the entourage effect" cast the cannabis plant's naturally occurring aromatic compounds in a new light. Rosenthal, then Russo, effectively outlined the interaction and synergy of terpenes with THC and cannabinoids, which basically illustrates that each strain's unique "high" is mostly based on how those aromatic compounds alter THC's effect on us.

High Pressure Liquid Chromatography testing for terpenes is utilized throughout the cannabis industry and is accessible to almost any cultivator in legal states. Buyers may finally be moving away from overall THC potency being a dominant market force with this new awareness and interest in terpenes. The typical cannabis flower tends to have about 1% terpene content, although there have been reports of some with 3%+. These levels will be increased by breeders.

With legal cultivation it will become easier for farmers to utilize classical breeding methods. For the best results a very large number of plants are cultivated from seed only a few are chosen, this process was very difficult to follow during prohibition. Even in the modern era many cannabis seed breeders still work in batches of selections of far fewer than 100. It's easy to imagine the potential if that number increased to a thousand or more!

Chem Dawg [photo by Nadim Sabella]

[all photos by Gracie Malley]

The other perhaps more potent element at play is the cannabis genome. Researchers already have a good working knowledge of the genomes various triggers for growth structure, phytochemical output (i.e. cannabinoids and terpenes) and many other traits. The threat of GMO cannabis seems to spook many from supporting these genome studies but in truth what we stand to gain is worth setting those fears aside. One of the more exciting prospects is the use of the principles of SMART (Selection with Markers Advanced Reproductive Technologies) breeding — using the knowledge of the genome to achieve the full potential of the plant, but breeding with traditional methods. It offers us our best hope in dealing with one of the most challenging issues that we face as the scale of cannabis cultivation grows larger; mold and insects or more specifically the pesticides used to fight them. By following the plant's genomic map and breeding towards increased resistance, we would be able to step closer to fighting infections with out the use of chemicals.

Pest resistance is not the only benefit that SMART breeding offers, future strains will be intricately and immaculately crafted works of scientific art. Imagine the ability to choose preferred cannabinoid content, terpene set and growth style like a chef in a kitchen with any ingredient at your disposal. By utilizing landrace, older strains, and feral strains the SMART breeder will be able to adeptly reach into nature's toolbox for the right genes.

I hope that the cultivator who is wondering what he/she should grow next can sense the pulsing current that carries cannabis persistently onward: newness. Finding it yourself is as easy as this: grow seeds and pick your favorites, retain males, check them for potency, yield, etc. and collect their pollen to make your own hybrids.

Finally, use quality lab testing to identify your most unique plants. The recipe for the next "elite strain" in your local marketplace — if not the entire world — is truly that easy.

Gorilla Glue #4 [photo by Justin McIvor]

The Heavy Hitters of Hash

Five strains favored by the world's top hashishins.

By Nikka T

Over the last 15 years, I have had the blessing of extracting over 5,000 unique cannabis varietals to create hash. There are many things to consider during the process of concentrating cannabis, but the most easily controlled, pertinent variable to focus on is the strain of cannabis from which the starting material originated. In that regard, I'd like to introduce you to my all time favorite top five hash making strains: Gorilla Glue #4 (GG#4), OG Kush, NYC Diesel (NYC D), Harlequin and The White. These strains and their offshoots all produce unique, yet oddly similar, chemical-like gasoline terpene profiles. Each also contains a broad spectrum of cannabinoids, while remaining stunning trichome producers. To achieve greatness these trichomes, or resin glands containing potent amounts of THC, should be light reflective clear, gold and amber colors and have oily, greasy, sticky and stable consistencies.

Favored strains for hash makers contain broad and high cannabinoid test results coupled with an overpoweringly strong and unique terpene profile that stings the user's nose and remains on the palette for hours.

Gorilla Glue #4

Gorilla Glue #4, also known as GG#4, was bread by Josey Wales and RJ and has gained extreme popularity within the last few years due to its large yields and unique chemical profile. The strain was an accidental miracle as the story goes. A Chem Sister plant's bottom branches turned into a hermaphrodite, which hit a Sour Dub plant, and four seeds were created. It took 3-4 months to get these seeds back and, once attained, only one seed popped creating what the breeders called Cha Ching. This plant turned into a hermaphrodite as well and created seven seeds with a Chocolate Diesel plant. From these seven seeds, Gorilla Glue was born. GG#1 and GG#5 will be released in 2016, but the GG#4 was spread

Gorilla Glue #4 [photo by Justin McIvor]

far and wide and has now won a plethora of Cannabis Cups and awards. The lineage of Chem Sister x Sour Dub x Chocolate Diesel created this monster known for its Gorilla Glue adhesive-like stickiness that leaves a binding residue on your fingers for quite a duration after merely touching a bud.

The plant itself grows fairly bushy and stalky in vegetative structure with very little stretching in flower. Medium sized trichome-covered dense buds line the thin branches and smell loudly of petrol fuel, anise, chocolate, lemon and pine. She finishes in about 60-65 days producing 1-2 pounds per light and leans on the clear-headed, yet strong, Sativa effect.

The Gorilla Glue #4 has caught my attention as a hash maker due to the bounty of trichomes that it produces alongside its pervasive and pungent terpene profile and extremely sticky amber gold colored resin.

Every single time we have processed the GG#4, no matter who the grower has been; we have hit over 18% yields of hashish from mere sugar leaves. Not only that, but the plant produces beautiful transparent gold oil in a multitude of gland sizes from 160 micron all the way down to the 25 micron, which is extremely appealing to the end user and extractor. In concentrate form, this specimen smells and tastes even stronger of gasoline and deep earthy chocolate. Potency numbers upwards of 89% THC put this strain at the top of my list.

OG Kush

Although the exact lineage of OG Kush is still unknown, many — including one of the original holders of the strain, Josh D — believe it to be an Indica consisting of Thai, Afghani and Hashplant origins. OG Kush is a lanky, low yielding plant and is well known as one of the more difficult plants to cultivate. It produces smaller, dense, extremely pungent, crystal covered, large calyx protruding nuggets after taking 60-70 days in flower. Acknowledged for being one of the more potent, high-THC testing strains, this specimen has an undeniable aroma and taste of lemons, fuel, pine, anise, a freshly opened tennis ball container, earth and sugar. Now, OG Kush, also known as Original Kush due to its many successors, is one of the most sought after strains to date. It has been renamed a multitude of times after being grown in different environments, regions and mediums over the years, changing its appearance slightly. It has also been bred with just about everything imaginable, multiplying its recognition via award-winning crosses like Girl Scout Cookies and Kosher Kush just to name a few. The rarity and hype surrounding OG Kush has lead it to be one of the leaders in the elite clone-only movement, creating phenotypes such as SFV OG (the soil grown version that has retained its notoriety in Southern California and beyond as one of the best), Ghost OG (spread to Oregon via the forums by OrgnKid) and many more such as '92 OG, Illuminati OG, Club 33, Tahoe OG and True OG.

OG Kush [photo by Nadim Sabella]

OG Kush Concentrate [photo by Nadim Sabella]

Another reason for OG Kush's intriguing nature is the mysterious stories and rumors surrounding its genuine origin. As far as the story goes, a grower by the name of Josh D, originally from New Jersey, moved to the San Fernando Valley in 1987 to pursue his passion of growing cannabis. Later, in 1996, a friend by the name of Matt, aka Bubba, moved from Orlando, Florida to Silverlake, California to join Josh D as roommates. Bubba would tell stories to Josh of this super unique strain originating in Gainesville, Florida, that they were previously growing, which nobody had seen before on the West Coast. Eventually, after Josh saw college photos of this hyped strain and pleaded to see it in person, Bubba finally transported cuttings out to their house in Southern California. Josh, wanting to soak up as much knowledge as he could attain, eagerly learned to clone at this point, and thus OG Kush and Bubba Kush, as well as another variety that was eventually phased out, were preserved and eventually spread from their tight circle in Los Angeles throughout the world.

Most of the genuine OG Kush varieties and phenotypes produce a very stout trichome structures often show short, thick, and oil-filled stalks with huge bulbous, almost bursting, gooey golden to amber glandular heads just begging to be decapitated. These gorgeous trichomes, once concentrated, exude that powerful gasoline, lemon, pine, earthy, anise, skunky funk of a terpene profile, making some believe the OG's origin may have some Chemdawg lineage. Although harder to perfect due to the extremely glue-like consistency, when processed in a cold and dry environment, OG Kush solventless hash is one of the most enjoyable coveted and desired smokes. Although it has smaller yields in flower and extract form, the complex piercing aroma and taste of fuel, anise and citrus, amazingly tacky consistency, transparent shiny golden color and strong body high, yet clear-headed effect the elite clone-only OG Kush and its offshoots will always keep this strain as a favorite hashish producing specimen.

NYC Diesel

NYC Diesel, also known as NYC D, was bred by the infamous Soma, of Soma's Sacred Seeds, by crossing the clone-only Sour Diesel — originally hyped on the East Coast for its large, pointy calyx structure, overpowering skunk and almost acrid petrol fuel aroma — with an Afghani/Hawaiian male. This resulted in a cerebral, talkative, yet all over body relaxing Sativa hybrid. For a Sativa leaning variety, this plant tends to grow a lot shorter and bushier, than its tall and lanky Sour Diesel predecessor and also finishes much faster (55-60 days), making this hybrid a great strain for the novice as well as the commercial grower. The intense terpenoid profile reflects that of a real skunk (earth and overpowering grapefruit notes while fruiting, in the jar, on the inhalation and exhale) plus, it lingers in the air with floral citrus aromas for hours after being consumed. Very much like the original Sour Diesel, and most of the Diesel crosses that have been popularized

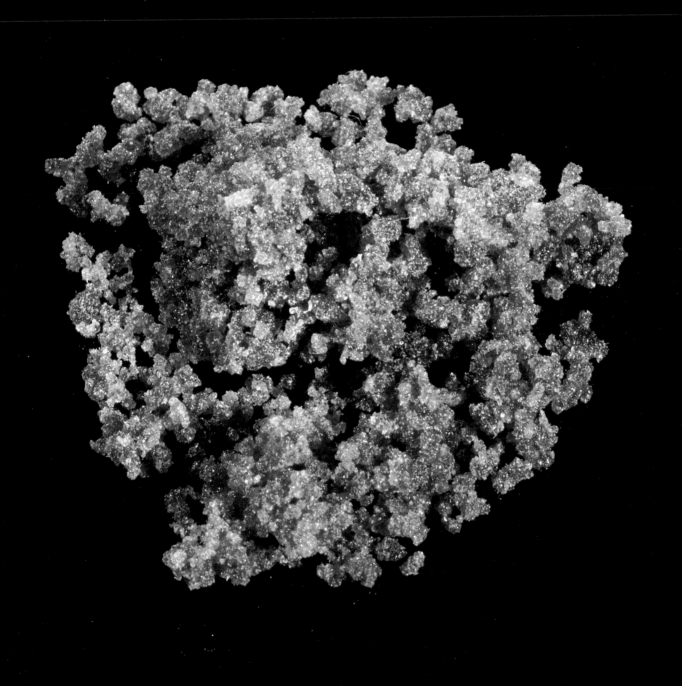

NYC Diesel [photo by Justin McIvor]

in the last 15-20 years, including Headband, Hong Kong, Sour OG, Sour Cheese, among many more, the NYC Diesel's buds tend to be larger and more conical in size, producing large yields of greasy, tacky, clear to gold tinted, transparent trichomes from the defining protrusive finger-like calyxes.

The NYC D, like the Sour Diesel, often displays a pink to magenta hue, not only in the leaves, but also on the actual buds themselves. In concentrated form, the NYC D's trichomes tend to yield fairly well, especially in the upper 70-160 micron silkscreen separating bags, due to the larger glandular heads containing the valuable terpeninoids, cannabinoids and that preferred greasy texture. Not only have we yielded over 2 pounds per light on a lot of the Diesel varieties, we often yield over 20% in extract as well. NYC Diesel tends to produce a very eye catching and appealing clear to light golden colored hashish, with an extremely viscous and sticky consistency, plus the smell of grapefruits, fuel and skunk so strong that your neighbors a few blocks away will smell the funk. A lot of Sativas that we process lean on a waxy, more granular consistency due to their innate nature of having longer, narrow, less oil filled trichome stalks, with smaller heads containing a larger wax-to-oil ratio than is generally desired by hash makers.

NYC Diesel is a rare and unique plant that grows like an Indica, produces beautiful swollen trichomes resembling those of common Indicas, and yet has the effect of a strong Sativa hybrid. This is an incredible strain with amazingly pungent citrus notes of grapefruits, tangerines, oranges and skunk and results in an uplifting joyous high.

Harlequin

Harlequin, known for its high CBD content, pain relief, as well as its anti-inflammatory and anti-seizure properties, is a Sativa dominant landrace strain that was only really discovered and preserved, like most other CBD strains, due to the demand of cannabinoid potency testing within the legal regulated industry.

Harlequin's roots are said to have come from Columbian Gold, a Nepali Indica, Thai, and Swiss landrace strains. Bill Althouse, a very humble healer and cannabis activist, has been credited with spreading this particular cutting of high CBD Harlequin to patients in need around the globe, absolutely for free. The strain itself, grows very wispy and delicate in nature, but if pruned and topped correctly, can form a wide, self-supporting, short and bushy gorgeous plant with thin Sativa-like leaves and extremely large forearm sized buds. Flower times range between 45 days (to keep the psychoactive THC levels lower) and 65 days producing over 1-2 pounds per light. The complex cherry cough syrup, black pepper, anise, earth and fuel aromas and tastes are very similar to a lot of CBD rich strains and is attributed to their healing characteristics via the pronounced terpene Beta-Caryophyllene combined with the entourage effect of other cannabinoids aiding in a plethora of ailments. Testing consistently around 14% CBD to 7% THC makes this 2:1 ratio strain an extremely viable medicinal resource.

Once the trichomes have been extracted from the Harlequin plant, a beautiful transparent amber colored, oily, yet sticky consistency, and pervasive cherry, pepper, anise, earth, and gasoline aroma is revealed exponentially stronger than the original flower it came from. Along with the medicinal CBD effects of this strain, especially in concentrate form hitting numbers as high as 60% CBD to 39% THC, Harlequin produces a large yield of perfectly tinted, extremely sought after, viscous oil, collecting large trichome heads in the 70-120 micron range.

The White

Originating in the Orlando, Florida region, this extremely rare and unique Indica dominating clone-only strain dubbed The White, has a completely unknown lineage. Its popularity started underground in Florida and remained coveted in the state for quite some time, until a man by the name of Krome popularized it by spreading the cut via IC Mag and and various other online private grower forums that took the underground breeding and cultivating market by storm in the early to late 2000s.

Like many strains, this particular variety popped up on the East Coast scene as a few different names including the Supes or Super Danks, Triangled and then later called That White and finally The White due to its popularity within Krome's small circle. The White is not a very large producing strain, but what makes it truly stand out is the outrageous coverage of white trichomes protruding from this dense, round shaped bud, creating the outward appearance of a white colored flower through and through. The White is a very strong and hardy strain in structure, has an extremely strong full body relaxation effect and a terpene profile that will rival the best. Some say the smell can often times be a bit subdued and bland, but when grown correctly, aromas of diesel fuel, earth, pine and sugary lemons pervade the nostrils.

The biggest reasons I have included The White in the best hash strains category is as follows. We haven't yielded less than 20% melty hashish from sugar leaf material, no matter whom the starting material was grown by, 90% of which tends to be full melt intact heads, the color has always remained a transparent fluorescent orange to gold, the consistency is extremely stable due to the clean separation from the plant material, the terpene profile is escalated tenfold from the flower itself revealing more deep complex gasoline, lemon, earth, and pine tree scents. Hitting THC levels in the 90s, and never failing me, The White will remain in my stable for quite some time.

The White [photo by NadimSabella]

Stigmas [photo by Professor P]

Becoming a Breeder: The History of Dynasty Genetics

Patient-based breeding program helps others to heal through cannabis.

By Professor P

It was the summer of 1996 in the sweltering heat of upstate New York when I decided it was time for a change. Over night, I gathered my seed collection along with other essentials and began the westward journey. At 18 years old and 2,800 miles away from home, the change was more intimidating than expected, but I soon became engaged with like-minded folks.

I began the next chapter in this new stage of life after getting a job and settling into an apartment on Mt. Hood in Oregon. It took several paychecks to get all the growing gear needed, but I eventually got lights, fans, soil and other basic supplies for the initial indoor seed crop.

I'll never forget germinating my first seeds and the joy they brought when I experienced them grow under the climate that I had personally provided. Cannabis and I had a very special relationship from the get-go and I saw that there was much more potential than I ever imagined. My path becoming a cannabis grower/breeder was unique from many others in that I made my own seeds and grew them out before I ever grew a clone, with no intentions other than my own self-sufficiency.

After a few successful seed crops, I had no choice but to produce more seeds before the original supply was completely diminished. With this process, came the determination to create the best strains for growers and friends in my community. I observed cannabis as a cure firsthand though a close friend who suffered from epilepsy. I only had one strain that effectively for suppressed his seizures. That was the one I focused on during the first few years of breeding. Witnessing those moments of healing, I began to understand the amazing potential of cannabis' medicinal qualities and I made sure to hold onto certain traits just for that.

My love and determination to craft the best medicine grew stronger with every plant I grew. Living within a booming medical cannabis community it was common occurrence to come across local and elite clones such as the Oregon Afghani, Oregon Blueberry (Purple Thai), Skunk #1, The Cough, Snowbud, original NL#1, NL#5, Strawberry fields, Purple Erkle, Trinity, Arcata Trainwreck, Chem D, Jasmine, Dogshit, NW Pineapple, William's Wonder, Blue Magoo and many more. I learned more than I could have ever imagined by growing and breeding many of these strains. Some don't breed as well as you might think, while others breed true. The learning process was sometimes agonizing, but that made the success sweeter.

In the late '90s and 2000s, I expanded my seed collection to several hundred strains including selections from South Africa, Afghanistan, Mexico, California, Oregon, Canada, The Netherlands and more. Seed collecting became more than a hobby, it was an obsession.

After a beautiful eight years living in the mountains with no computer or internet, I moved closer to the medical scene that I had been involved with over the last decade. Shortly after my move, I had a near death experience. My friend and I were hit head on by a drunk driver. Unable to walk for months, I discovered a profoundly greater understanding of medical cannabis and the looming dangers of pharmaceutical pain medicine. This was one of the biggest moments of clarity that I've had in my entire life.

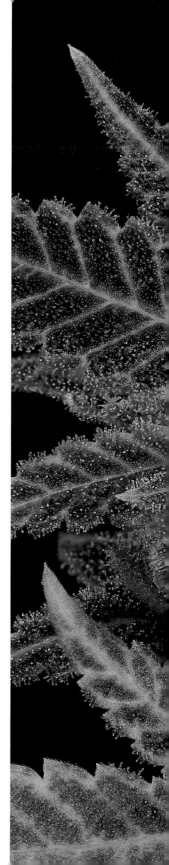

Starduster [photo by Professor P]

Starduster Greenhouse [photo by Professor P]

Over the following few years I helped many patients by donating seeds, clones, buds, hash and edibles. Growing and being a caregiver broadened my understanding of ailment specific plants and their healing capabilities. After my accident I became stricter on the parental selection process and usually, choose medicinal values over other traits. This is done using patient feedback.

In the past I had been extremely partial to Sativas and Sativa hybrids, but my work expanded into the Indica realm after one of my best friends lost his leg in a snowboarding accident. The most effective Indica for his post surgery was Oregon Afghani, but it didn't yield enough and had a long veg time. A mutual friend approached me about making a Oregon Afghani hybrids for our friend. We ended up making several Indica hybrids using the clone provided and they were deemed lifesavers. Working side-by-side is motivating. Our extended helps adapt strains. I'm very grateful to work with masterful growers. Passion for patients stories over profits has been our moral compass. Our patients make us strive to be better growers, breeders and most of all, better people.

After implementing this "function over fashion" breeding program, we've developed strains test with anomalies previously unseen. Ms. Universe strain is used to hybrids with high percentages of CBG and terpenes. This does not translate to THC percentage, but has other medical qualities.

Finally, I believe that the more aware we are about the effects of cannabis on our body, the more we will become in tune with our inner being and nature itself.

Platinum Huckleberry Cookies Trichrome [photo by Professor P]

Women Grow Leadership Summit [photo by Christian Purdie]

Women in Weed

The female force behind cannabis continues to rise.

By Susan Cohen

In August 2015, Newsweek published a cover story about women working in weed. "Though the industry is still predominantly male and employment statistics are somewhat vaporous," writer Gogo Lidz wrote, "the power and influence of women are, by all signs, on the upswing."

That may have been a surprising fact for readers of the magazine, but for the women who actually hold those jobs, it's their reality. If women could play crucial roles in the fight against cannabis prohibition, penning legislation and rallying others for the cause, why shouldn't they get a fair stake in this booming billion-dollar industry?

In fact, women have been essential to cannabis' massive growth over the past few years.

They've established all sorts of cannabusinesses, from dispensaries and product lines to staffing agencies and law firms. According to a 2015 Marijuana Business Daily survey, women hold 36 percent of all executive-level jobs at cannabis businesses, compared to 22 percent of their peers outside of the industry.

Women know what women need, and a number of these new businesses were created to reflect the needs of female cannabis enthusiasts. AnnaBis and Erbanna both design fashionable purses and bags that control product odor, while Floria Pleasure and similar companies focus on female sexual pleasure. Even Whoopi Goldberg has entered the market — along with Om Edibles founder Maya Elisabeth, Goldberg now produces a line of edibles and topicals designed with women in mind, including a cannabis rub that eases menstrual pain.

However, while plenty of new opportunities have been created for women, it hasn't happened without risk. In September 2016, the Center for Investigative Reporting shared accounts from female trimmers who reported sexual abuse and human trafficking on pot farms in

women
GROW
ultivating cannabis entrepreneu
Connect • Educate • Empower

WomenGrow.com f ✕ in

Grown&
Extracted&
Tested&
Baked&
Purchase
Enjoyed
By Wom

WomenGrow.com

Women Grow Brand Table [photo by C Roese Rampl]

Jazmin Hupp, Co-founder of Women Grow [photo by C Roese Ramp]

California. Meanwhile, December Kennedy, the COO of a Denver greenhouse and lighting company, recalled in an essay on Canada's Cannabis Digest blog the time police officers and social workers arrived at her house. The mother of three ran an edible business at the time, and the group wanted to see her garden. After a quick inspection, the police officers raised no concerns and left, but Child Protective Services stayed behind. Then they opened a case on her.

Kennedy was lucky; she didn't lose her children, but many other mothers have. Despite this risk, women are still stepping up, and they have banded together publicly to endorse their needs and support each other. Founded by Jazmin Hupp and Jane West in Denver in 2014, Women Grow brings female entrepreneurs together through networking and other events. There are now 26 chapters in 14 states and the District of Columbia, plus another three in Canada, and it's not the only group that advocates for women legally and economically. Others include NORML's Women's Alliance, Washington's Women in Weed, and California's Synchronicity Sisters

Women have also become some of the most recognizable faces in cannabis. Marijuana use is a central component of the comedy of Broad City's Abbi Jacobson and Ilana Glazer, while High Maintenance co-creator Katja Blichfeld recently developed her Vimeo comedy series into a program for HBO. For her advocacy work and infamous televised F-bomb, Charlo Greene is one of the movement's top celebrities, though she currently faces significant jail time for operating her Alaska Cannabis Club. And Cheryl Shuman, the "Martha Stewart of Marijuana" and founder of the Beverly Hills Cannabis Club, has made efforts to appeal to a more traditional audience.

The cannabis industry that exists today essentially didn't five years ago. That makes it different from traditional trades: It isn't bogged down by generations of glass ceilings. As a result, the expanding field could be the first to include as many women as men — and women are making sure of that.

Lobbying Day [Photo by Women Grow]

Hash Cannolis [photo by Gracie Malley]

The Lost Art of Temple Balls

A hashish master and his student explore the preservation of resin.

By Frenchy Cannoli and The Dank Duchess

Charas, or hand rolled resin, is the original concentrate. Alluring and aromatic, charas was born from the first contact between humanity and the cannabis plant, as a layering of this sticky THC-rich substance is unavoidable when handling marijuana. Innovations in cannabis concentrates, utilizing newer methods of extraction such as BHO and CO_2, will soon transform the face of modern day medicine. Still, the act of gently rubbing cannabis flowers will remain the easiest and most effective method of collecting fresh resin from wild plants at the peak of their life cycle and creations like the Temple Ball will continue elevate what could be construed as mere collection into an artisanal craft.

The process of collecting live resin in the palm of one's hands is simple in its methodology, but challenging in its implementation. While no longer widely-practiced, this method remains the sole cannabis resin collecting methodology in tropical countries with a humid climates like Bhutan, Nepal and Northern India. To collect resin, take the fan leaves off the plant and gently caress the flowers between your palms using a light back-and-forth rubbing motion. Thoroughly clean your hands of any leaf material after each flower and start again until a layer of resin builds up on your palms and fingers. Then snap the substance off your hands and voila! You have created hand-pressed resin.

Hash Rolls [photo by Gracie Malley]

The feeling of the resin slowly collecting, plant after plant, is a unique tactile communion and an unbelievable olfactory experience.

There is an indescribable intimacy and closeness that is born from such a synergy, a communion that goes beyond the plant and connects to the terroir that gave birth to the magic. But the relation between a master gatherer and the resin also extends beyond the realm of collecting. In tropical countries preservation and aging are essential to quality and longevity. To this end, a Royal Nepalese Temple Ball was the ultimate manifestation of resin optimization and preservation. The origins of such a cutting-edge approach to packaging and long-term conservation may never be discovered, but the art should not be lost.

The Royal Nepalese Temple Balls were stuff of mythology already in the late '70s and early '80s, a fairytale for many and the Holy Grail of concentrates for a few. Imagine a sphere of resin hand pressed to an absolutely unflawed dark and hard surface polished to a mirror-like quality — a ball that resembles more of a glossy stone or black marble rather than resin. Picture an outside protective layer of resin fused into a crust so perfect that it can stand the depredations of time and nurture the aging evolution at its core. Visualize cracking open a 10-year-old resin ball like you would break a big egg, the center revealing itself slowly in all its glorious creaminess.

Envision a spicy tropical fruit cocktail with subtle earthy undertones taking over your olfactory senses as the resin breaks apart reluctantly exposing its dark red melted caviar like texture, the long contained aromas bursting out with an explosive force.

Imagine creating such a wonder!

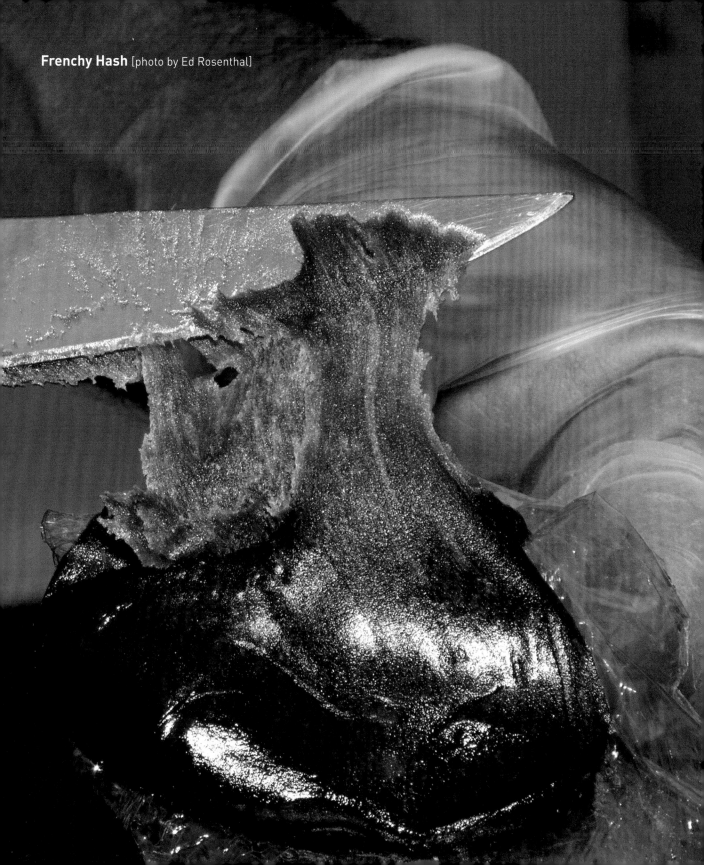

Frenchy Hash [photo by Ed Rosenthal]

Hash Cannolis [photo by Gracie Malley]

Hash [photo by Gracie Malley]

A Modern Interpretation of the Temple Ball by The Dank Duchess

Frenchy's eyes twinkle as he tells me stories of hashish around the world and indulges in nostalgia as he details his rich experiences of Malana cream and Nepalese Temple Balls. With true Nepalese Temple Balls no longer anywhere to be found, I can only imagine the creamy inner goodness of which he spoke so lovingly. Perhaps I can create something for which he will be equally fond.

I start with well-dried Jah Goo resin; known for its potency and intense flavor. I place a thick glass bottle filled with boiling water on top of the resin, which has been wrapped in a plastic steam pouch. It begins to melt almost immediately. I can see the yellowish color of the resin darkening as the heads fuse together, as if becoming more saturated with life. It is tempting to hurry the process along, but the resin moves at its own pace. I simply hold the glass bottle as a guide, slowly rolling it forward and turn the resin, gently rolling the bottle away from me. Flipping it, I see that the resin has begun to melt on the other side as the heat has easily penetrated. I pick it up and squeeze softly through the plastic. It feels completely melted through. With a swift flick of the wrist, I snap the plastic open and the aroma of plump, ripe, wild berries wafts upwards. Jah Goo's dirty blonde color has transformed into golden amber.

The resin gleams with oil and expectation. Its satiny surface reflects like a mirror. Folding the resin in half, I snap one side of the plastic; and then the other, grabbing the resin simultaneously. Folding it one more time, I notice that the interior of the resin has retained its pre-melted color and rather than holding firm, it separates like bread dough that is not fully kneaded. I place the resin between the plastic and begin rolling the bottle again. This time, the resin reacts even faster, spreading across the plastic as if running from the heat.

I begin shaping the Jah Goo hashish when the entire resin softens like warm marshmallows. Removing it from the plastic, I fold the resin in my hand twice, feeling the warmth radiating through my palm, and enjoying the fruity and spicy scents that tickle my nostrils. Using my first two fingers and my thumb, I begin shaping the resin into a cube. I squeeze firmly and squeeze again, before rotating, repeating the cycle until there are no wrinkles on the surface. Squeezing has removed the air pockets, leaving a dense block with sunken concave sides. I start pushing the corners in slightly, massaging the hashish into a juicy dumpling. I clasp my palms together and slowly roll my palms back and forth.

Finger Hash Balls - Khandwa India 1981 [photo by Ed Rosenthal]

After getting a rhythm going I pause every 10 seconds to look at the progress of the ball of hashish. It has become as smooth as an egg and a similar oval shape. I love touching the resin, but hashish must not be overworked. The hashish is almost ready, and I roll faster smaller circles as it cools in my hand. Finally, it is a shimmering ball of Jah Goo hashish. It has become a taut, shiny marble glimmering in the light.

Traditionaly Temple Balls were rolled on a ceramic plate creating an impenetrable crust. Now, it's often left with a stain sheer. I place the Temple Ball on a sheet of parchment paper. Oil continues to express itself more and it flattens on the bottom under its own weight. It is gorgeous and smells like sheer happiness. This is not the perfect Temple Ball of Frenchy's memory, but I hope with a bit of aging, my offering to him will warm his heart even more.

Temple Balls [photo by Ed Rosenthal]

Frenchy Cannoli [photo by Gracie Malley]

[photo by DevilsLettucePH]

Styling an Industry

DevilsLettucePH strives to show the full range of nature's beauty.

By Ellen Holland

For Randi, the photographer behind the lens at DevilsLettucePH, quality image samples come in the form of the freshest herb. Cannabis transported straight from the plant into her hands is best, she says, while laughing at the notion of receiving buds squished away inside of a bag. Self taught and driven by a desire to help others find the healing nature of a plant demonized for centuries, Randi now finds herself redefining the design aesthetics of the world's favorite botanical. Taking marijuana into the new legal era means applying upscale marketing techniques to the dankest strain selections. To achieve her images of cannabis, which bring the flower into an elevated retail space in much the same way as an advertising campaign for a beautiful bottle of perfume, Randi follows her inner intuition and only takes photographs that she herself enjoys.

"Time and time again, when I compromise for what others like the shot suffers," she says while offering advice for newcomers. "Stay true to your soul and what jives with you."

Sampling from the plethora of amazing genetics available in the San Francisco Bay Area, Randi's images playfully pull in other visual elements to enhance three main stars: beautiful, crystal-coated dried buds, marijuana plants in growth and concentrates with incredible structure and clarity. Ideally, the growers themselves provide Randi with their hand-picked selections.

"If you don't have access to a grower and can't grow your own, you can secure quality herbs at collectives," she says.

Randi explains that cannabis can be a difficult subject to shoot because of the complexity of its surface and stresses always keeping the subject in focus. In her work she attempts to target an interesting and dynamic depth of field, an aspect of photography she has developed since her early days of documenting her explorations in nature.

"I always liked to take pictures of nature or lines, but always at odd angles and creating a large DOF," she says. "This to me created a more dramatic effect and helped the photo become more than that, it became more than just an image captured in time, it became an artist's point of view, I was showing you how I saw the world."

For Randi the photographs have become her vehicle to spread the healing message about medical marijuana. Diagnosed with Lyme disease about 11 years ago, she turned to cannabis after being prescribed medications that only made things worse.

"Cannabis has made life tolerable, without it I wasn't very functional, always sick," she says. "Since cannabis was able to help me so much, I wanted to give back somehow and the only way I knew how was with photography. I saw a need — two years ago now — for more, in my mind, beautiful pictures. Images that would inspire others to change question how something so wonderful could be bad, something to help reverse the negative propaganda from the past."

With her husband's encouragement and artistic insight — Randi reveals he picks many of the strains she documents and was the first to initially bring her mouth-watering buds from Cannabis Cups— she branched out from her current career as a software engineer at a tech company into the world of marijuana photography. Now, she says, her heart is in every image.

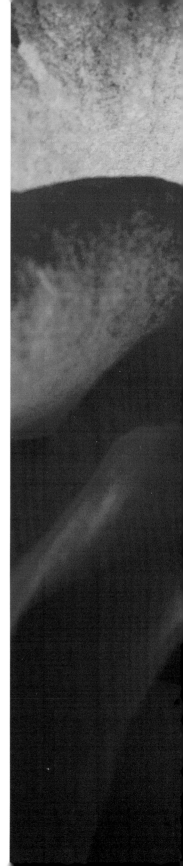

Rose with Shatter [photo by DevilsLettucePH]

Spider & Dragonfly [photo by DevilsLettucePH]

[photo by DevilsLettucePH]

"Corny yes, but when my heart isn't in it, it shows, so I tend to shoot with music."

In order to find diversity in her shots she often explores pairing cannabis with props.

"I get bored with the same old same old, so I like to change it up," Randi says. "I also like to find colors or props that speak to me about the stain or bud. Some of my favorites are crystals, rocks, wood and metal. Toys are a hoot and using other plants is fun too... it's really just what sings to me at the time."

"It can be more challenging working in a garden," she says, "it's the hardest to work directly in the environment." Still, she looks forward to the opportunity to document marijuana plants in different scenarios. She says "my home studio, while more controlled, can also be limiting as I am forced to work with what is around as opposed to exploring further unknowns the enviroment. It is more fun to have diverse subject matter."

As Randi develops her craft, she is optimistic that her images will help dispel negative stigmas that are still attached to the healing herb.

Cannabis Concentrates – Extraction's Resin Renaissance

The buzz around dabbing has fueled a whole new era of potent marijuana concentrates.

By Greg Zeman

Centuries after we've all toked our last, when hash historians of the future trace the evolution of whatever fantastic extracts they're ingesting, they will break everything down into the BD and AD eras – Before Dabs and After Dabs.

The social buzz around dabbing has done more than ignite new interest in the very old idea of extracting and concentrating cannabis compounds. This resin renaissance has created a new cultural framework for the categorization and consumption of concentrates.

The earthy spice and balanced effects of well-crafted "full spectrum" hash will never go out of style. But if you haven't waded past the icy shallows of traditional cold-water concentrates, there's an uncharted sea of exotic flavors and otherworldly sensations waiting for you to dive in and explore.

My "beat" is chasing the bleeding edge of a technological and cultural revolution in cannabis extraction. Thankfully sampling the most powerful concentrates on the planet while doing it.

As a result, I get regular opportunities to speak (and dab) with some of the finest minds in the industry, people who are changing the way the world approaches cannabis consumption. From elusive chemistry majors hell-bent on

Rosin after heat press [photo by Professor P]

perfecting the molecular isolation of liquid terpenes and THCA crystals, to self-taught closed loop artisans who've mastered the art of making shatter that tastes like sweet fruit and looks like clear golden glass.

California is awash in more dabbable extracts than I can ever hope to sample in a single lifetime, not that it stops me from trying. But it wasn't always like this. The first time I tried any sort of concentrate that wasn't bubble hash or traditional style hash oil was in the early '00s — whenever the Vapor Room in San Francisco first started selling CO_2 wax.

By that time I was smoking hash with my morning coffee, so one of the VR's budtenders suggested the wax on the day it arrived. He definitely warned me to take it easy on my first try, so my only excuse for what happened next is that I was young.

My plan, to the extent I ever had one in those days, was to try out the expensive wax stuff, ride my bike to the Red Vic and catch a movie. Eager to experience the full effect, I mashed a grape sized lump on top of a bowl of flowers and fired it up, taking the same sort of massive rip I normally did, completely ignoring the nice man's perfectly sensible advice.

The explosive coughing fit that followed lasted about five minutes, but the overwhelming physical and psychological effects lasted hours that felt like years. That night (which I spent alone, teetering on the razor edge of full-blown hallucination) wasn't entirely unpleasant, but suffice to say I never made it to the movie theater.

Rather than being put off by the mind-blowing experience, I was intrigued. I started buying wax instead of hash, but intuitively felt I wasn't utilizing it to its full potential.

I started experimenting with some early "wax pens," which made for more convenient use but never quite captured my fancy. Loading them was a

Shatter [photo by Steephill Labs]

Getting a dab ready with vblogger Coral Reefer.
[photos by Gracie Malley]

Heating the nail.

The shatter vaporizes on the hot nail.

Rosin [photo by Professor P]

pain, they had battery issues and, on balance, were an over-complicated buzzkill.

And then I remembered how we used to smoke hash when I was in high school — hot knifing. We'd heat up steel knives in the coils of an electric stove and press a chunk of hash between the eponymous hot knives, capturing the resulting smoke with a paper funnel or half of a two-liter plastic bottle.

I asked a few people about hot knifing wax and they told me all about dabbing. I went to a local headshop, where the manager was happy to sell me a cheap dab rig at a 500% markup, complete with an equally overpriced titanium nail and crummy butane torch.

That night I excitedly took my first dab in classic first dab fashion — off a red-hot titanium nail.

Fast forward to 2015. I'm taking a low-temp dab of 99.9% pure crystalline THCA rolled into clear gold shatter. I feather the carb cap on the quartz glass banger until I'm satisfied with the size of my hit and blow a billowing cloud of terpy vapor at the computer screen as my lungs heave a sigh of relief – my how times have changed.

So what's the difference between shatter and wax? What about crumble and sugar?

Putting aside rosin and CO2 extraction for the moment, almost all the wax, crumble, shatter and so on you find for sale in states with dispensaries is made using solvents like butane.

One common catchall term for these solvent-based extracts is "BHO," which stands for butane hash oil.

And while that term is imprecise, it's useful for explaining how wax and shatter can be so chemically similar yet have such radically different physical characteristics.

One easy way to clear up a lot of the confusion surrounding the different types of "non-hash" extracts is to think about concentrates like candy.

Most candy starts out as a simple blend of sugary syrups. The difference between a lollipop and a piece of taffy is largely the result of temperature during cooking. In the same way, the immediate result of running solvents through cannabis will always be some liquefied blend of extract and solvent, it's what comes next that determines its final state.

So the same run that gets purged in a vacuum oven to make a shatter can be whipped over heat to make a wax. But the core method of extraction remains the same.

Some people adore what they perceive as higher terp levels in wax, others are fascinated by the golden clarity and hard candy stability of a great shatter. Many of these differences are superficial and a matter of preference. But however the concentrates are made, dabs are part of cannabis' future. Those of us who have tasted and felt the incomparable power of a good dab know there is no substitute.

Dab Tool [photo by DevilsLettucePH]

710 Glossary

By Greg Zeman

At the dawn of California's cannabis culture, the best hash still had to cross an ocean to get here and the dankest herb came tied to a stick. The basic fundaments of cannabis culture, never mind the finer points of effective cultivation, were spread predominantly by word of mouth.

Cannabis publishing pioneers like Ed Rosenthal and Tom Forcade started making vital information available on a broader scale and the advent of the internet only accelerated this sharing of knowledge. Now of course, the aspiring cultivator or hash maker has no shortage of resources available to them and California has seen more than one cultural and technological renaissance in cannabis.

But the broad embrace of critical extracts and the ascendance of "dab culture" represent a seismic shit in the cannabis landscape. And as with the birth of any subculture, a new set of words and ideas have popped up that may leave the uninitiated puzzled.

So for your edification, here's a rundown of some potentially unfamiliar terms you're bound to encounter when exploring the world of critical extracts.

710:

This is just "OIL" spelled upside down using Arabic numerals, but it's a core symbol of the movement in the same way 420 is symbolic of traditional cannabis culture. Oil, sometimes stylized as "errl," is often used as a sort of catch-all term for critical extracts, although this is becoming less common as enthusiasts become more fluent with the different categories of concentrate. If you see 710 used in the name of products and events, that means they're geared towards dab and extract aficionados.

CO2 Extraction:

The same critical extraction process as BHO, but liquified CO_2 is used in place of liquid solvents like butane and propane, resulting in an end product totally free of residual solvents (see residual solvents). Most "vape pens" that utilize pre-filled cartridges use CO_2 extract. Historically the product has been harsher and less flavorful than solvent-based methods, leading to re-flavoring and the addition of artificial terpenes, but advancements in technique mean some newer products are on par with BHO in terms of flavor.

Open Blasting:

This is a low-tech extraction method that doesn't utilize a closed loop system (see closed loop extraction) and prioritizes speed over safety. Trim or flowers are stuffed into a glass or metal tube and liquid solvent is run through the tube. What comes out the other end is a highly volatile blend of evaporating solvent and the desired cannabinoids. When this process is done indoors and/or by careless people who don't know what they're doing, the result can be a deadly explosion. When done properly, the risk is minimal but still present.

Closed loop extraction:

This is the preferred method for extracting critical extract. The blasting process is still at play (confined space + weed + solvent) but it is a closed system using a series of tubes, columns and chambers to contain the entire extraction process. The benefits are greater control over your solvent and next to no risk of explosion if used correctly.

Purge:

The purge process is absolutely crucial to safe product. It generally utilizes heat and agitation (or vacuum treatment) to remove the solvents used for extraction. The method used will also determine if the end product is a shatter or a wax concentrate. This process, when done properly, is what makes solvent-based extracts safe for consumption.

Residual Solvent:

Butane, propane and other solvents typically used in critical extraction are considered safe for human consumption in small doses and make regular appearances in packaged and processed foods. The purge process reduces the level of solvents in the final product to safe levels, but there will always be some solvent left in the final product, which is why some users prefer CO_2, rosin and other solvent-free methods. Lab testing can ensure that the solvents still left after the purge (residuals) are below safety thresholds.

BHO:

This is an acronym for butane hash oil, which is sometimes used as an imprecise general term for any solvent-based extract. BHO only refers to the solvent used in the extraction process and the end product can take a variety of physical forms depending on how it's processed and purged.

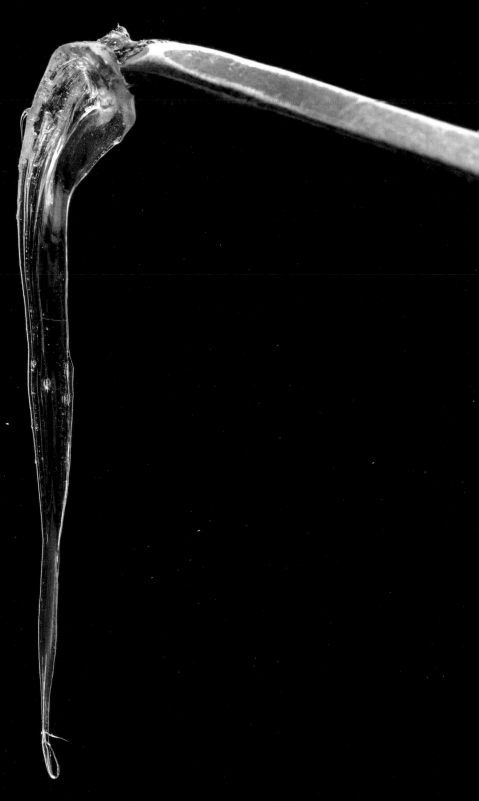

BHO Dab [photo by Professor P]

Seedling [photo by Professor P]

Pheno-hunting: Creating the World's Greatest Strains

Invent new types of marijuana by following the advice of expert breeders.

By Harry Resin

The dankest new marijuana strains sometimes originate as happy accidents, mutations or exceptions that pop up during a breeder's journey through an experience known as pheno-hunting. Abbreviated from the word phenotype, a scientific term that essentially means different versions of the same strain, when pheno-hunting cannabis breeders search through vast numbers of plants in order to find one with a number of exceptional qualities such as resin production, potency, weight, aroma and flavor.

Different phenotypes of marijuana can be compared to sisters who share a number of family characterics, but are not exactly alike. Depending on the breeder, marijuana seeds will either be stable and express a mostly uniform outcome or, if the strain was not stabilized, cultivators will end up with myriad of phenotypic possibilities. Breeding new strains of marijuana is based on hereditary inheritance that originates from the chromosomes of the parent donors. Like humans, cannabis plants reproduce using male and female aspects and inherit traits from both parents.

Since the beginning of time, man has selected plants based on varying traits, always looking to increase or better his crops from the previous season. This is also the case with cannabis. Marijuana breeders throughout time have selected strains based on characteristics such as flavor, color, or yields. When a breeder starts a project it is usually with a specific goal in mind, but one sought-after quality common to most new strains is vigor. Generally seeds will have more vigor than clones, especially certain hybrids, as they will combine the best traits of both donors or parents, thus inheriting a built-in strength.

It's always interesting to sit down and talk to different breeders because we all, for the most part, work the in the same way, selecting both males and females and then experimenting with the outcomes of crossing the various building blocks. Modern science has now also entered the fray allowing for marker-assisted breeding projects that use genetic sequencing to discover various traits, or markers, that can be used in advanced forms of breeding.

In order to gain insights into the process of creating exceptional unique types of marijuana, I spoke with a sampling of cannabis breeders working all over the world. They included Mat from The House of the Great Gardener, Big Buddha and Milo from Big Buddha Seeds, PC from Purple Caper Seeds, Jolly Mon from Loompa Farms and Jen Norstar from NorStar Genetics/TGA Genetics.

To understand what my fellow breeders have in mind when they start a project I began by asking about their goals. Mat from The House of the Great Gardener says that when he starts his selection process his goals often vary. He typically looks for strong vegetative growth as a start, vigor is, of course, very important. Great smell is also an essential trait he looks for both during the vegetative and flowering stages. Most good strains will already start to give off scent while they are still vegging and new cultivators should pay close attention to both their eyes and nose in the growroom. While the optimal aromas will vary from project to project, marijuana bursting with fruit fragrances such as strawberry, guava and cherry, are currently showing wide appeal.

Big Buddha, from Big Buddha Seeds, emphasizes having a clear goal when starting a cannabis breeding project. Examples of a breeder's intent could include creating a quick-flowering Sativa, a high-yielding Indica or focusing on other key attributes such as mold resistance or general plant size.

[photo by Purple Caper Seeds]

Tangie [photo by Justin McIvor]

Breeder Lawrence Ringo at work [photo by Sohum Seeds]

Jolly Mon from Loompa Farms seconded Big Buddha's advice by saying prospective breeders should establish a direct objective and have an endpoint in mind when starting a project. Establishing clear intent was also stressed by Jen Norstar, who said it's important to keep improving your breeding stock while also staying focused. When you start a project, pick out where you would like to end up, either with something in a particular terpene bracket or with a certain structure or yield. As Milo from Big Buddha explains, having a goal is key to establishing a well thought out breeding program, otherwise you can get bogged down in the work and the multitude of crosses.

Mat also emphasizes that when creating medical strains it's important to conduct outreach with patients in order to test the efficacy of the medicine. He works in Canada with several compassion clubs that also have access to testing, an aspect of modern day marijuana breeding which he finds super useful. With lab-tested results breeders can compare the different chemotypes (the chemical profile of a bud) of their plants. Each chemotype is like a fingerprint and is unique, providing invaluable data towards breeding strain selections. Admittedly, I have found the same usefulness in working with a lab here in the states after moving from Europe. Certain labs, such as the Berkeley-based Steep Hill Labs, can sex test and chemotype leaf samples from plants that are only three weeks old, speeding up the breeding process by months and months. Now, instead of a two month long selection process, results are back in a few days and you can know whether or not your plant is male or female and understand its exact chemical makeup.

As Buddha's seeds are for the most part feminized, or derived from strains that have been bred to produce only female seeds, he explained that knowing how female and males react is also an important aspect to his breeding program. An advanced breeding technique involves using female plants to make male pollen. When using this technique it's very important to select the right female as certain females are sterile while others are incredibly generous pollen donors. Feminized seeds are generally made by spraying STS, a chemical component used in the photography industry, on a female plant. The stress of this chemical will cause male pollen sacks to form on the female flowers. Pollen created in this way will only contain female chromosomes and therefore, when used to pollinate other females, will produce female-only seeds. Having a stable supply of female plants that exhibit large amounts of female-only pollen is quite desirable to breeders making feminized seeds.

In standard marijuana breeding knowing your males and having a secure group with predictable traits is also crucial as it helps to develop an understanding surrounding recessive and non-recessive genes. In layman terms, this means understanding whether or not you have a dominant male. As a breeder, I've seen elusive non-dominant males that impart none of their characteristics as they have incredibly dormant genes. In this circumstance your female offspring will be very similar to their mother, if the mother is dominant, a result that can be useful when working with exclusive clone-only strains. By re-invigorating them in this way, they will have an increased strength — a quality which will often mean larger yields and healthier plants — but they will also retain all of the desirable traits from the female plant.

All those I spoke with have an extreme love for cannabis and encourage new breeders to develop their own passion for marijuana through discovery by crossing different strains and eternally questing for new exciting outcomes. Mat from the House of the Great Gardener suggested growing out as many strains as possible in order to expose the multitude of variations from many phenotypes, then honing in on the specific traits that you desire. PC agreed, stating that breeding cannabis is a numbers game, so one should start with as many seeds as possible.

The more beans you crack the better your odds are of finding something amazing. While the number of seeds you start out with should be high, Jolly Mon from Loompa Farms suggested limiting yourself to only a few strains at a time.

Several of the breeders including Buddha and Jen suggested that the final key to success is testing. Let others grow out your new seeds and lab test the bud that comes from the finished selections. All of the breeders agreed that it's important to seek out the opinions of fellow breeders and friends. How was that new strain really? What was the best phenotype? It's essential to surround yourself with a good group of growers and smokers. Buddha said his number one tip is to listen to the people around you, the dispensaries buying your bud, fellow breeders and smokers. Go out there once you've made a few new crosses and test what you've made, it's really valuable to get feedback. You will see there will always be one or two super special strains in the mix, those few amazing keepers. As a breeder these strains will become prized trophies gained in your own pheno-hunt, success stories in the never ending pursuit of better bud.

Loud Scout [photo by Loud Seeds]

Sour C Train CBD Hash [photo by Nadim Sabella]

The Golden Age of CBD

By Jimi Devine

In the 2000s strains heavy in cannabidiol, or CBD, took root in the hills of California. While the cannabinoid had first been isolated in 1963, thanks to the budding analytical lab scene breeders and growers in modern times now began to take note of its wide range of healing applications. Once bred out of much of the California's cannabis gene pool due to a push to develop THC-laden heavy hitting strains, CBD was soon breed it back in for its therapeutic benefits, which at the time still remained largely unknown.

It took a nonprofit organization and a mainstream television special to bring CBD to the masses, but today there are more and more CBD-strains than ever before.

In 1964, a year prior to his discovery of tetrahydrocannabinol, or THC, Israeli chemist Raphael Mechoulam discovered the chemical structure of CBD. However, the cannabinoid failed to spur research interest through The Summer of Love, and for good reason — Mechoulam published a paper in 1970 declaring CBD didn't have any additional psychoactive effects based off feeding hash to monkeys. This period marked the first wave of CBD research and much of it was focused around its relationship with THC in the body and the first rumblings of CBD's now famed anti-epileptic properties.

With the earliest of trials showing an extremely low toxicity and no side effects, it was on to human testing. In the original test group of 16 healthy folks all showed no ill-effects from the CBD, and all the subjects slept well the first week. Now it was on to eight sick people suffering from epilepsy who were given daily doses of up to 300 mg of CBD and monitored monthly for up to four and a half months. The results were promising, four of the participants had almost no seizures for the length of the research and three of the other participants showed improvement in their symptoms, with the last patient seeing no effect on their condition. The CBD used in the study was an isolate very similar to what we see returning to dispensary shelves today and was produced by Mechoulam from Lebanese hash provided by Israeli Police. Despite these promising results, interest in the CBD research field continued to fade for decades.

CBD Standouts

There are many product options out there for those seeking the healing properties of CBD. Here are some healthy, smokeless applications of this incredible cannabinoid.

Incredibles - Black Cherry CBD Chocolate Bar

This delicious CBD-offering has received a Cannabis Cup award for the best CBD edible and tastes amazing. Crafted with milk chocolate and dried cherries these bars also offer 50 milligrams of CBD and 50 milligrams of THC. The dosing ratio on this product, make it a good choice to relieve chronic pain.

Kushy Punch - CBD Gummy

Kushy punch delivers potent effective medication throughout its line of edible gummy products, but the CBD gummy is particularly effective for reducing social anxiety. These gummies come with 30 milligrams of CBD and 60 milligrams of THC.

Moxie Meds - CBD Tinctures

These CBD tinctures from Moxie Meds come in both 1:1 and 1:4 CBD to THC ratios and can be used for a variety of ailments. The 1:1 ratio works well for women seeking relief from the cramps and pains associated with their monthly hormonal cycle and the 1:4 ratio is ideal for minimizing stress and tension, providing assistance with issues such as anxiety and emotional stress. These tinctures are made with full plant extract and MCT oil, a form of saturated fatty acid that has shown many of its own health benefits including antioxidant properties.

Wild Earth Naturals - CBD WATER

Wild Earth Naturals' CBD Infused Water uses a "revolutionary" nano-sizing process to maximize bioavailability – delivering ion-sized nutrients that are encased within water molecules which can easily move through cell membranes. Each 16.9 ounce bottle of CBD Water has one serving of Cannabidiol.

Hermetic Botanicals - Calming Companion Chews

Is your cuddly companion troubled by loud noises? Are fireworks and thunderstorms a source of anxiety and nervousness? Hermetic Botanicals brings you an innovative solution. Calming Companion Chews encapsulate all of the amazing benefits of hemp in a delightful gummy for your pets to enjoy.

Cats, dogs and even horses have found tremendous benefits from these tasty treats. It'll be love at first bite!

SENSI CHEW - CBD PLATINUM

Sensi Chew CBD Platinum 100 mg G is for patients with chronic pain who want relief with no THC psychoactive effects. It contains a total 100 mg of concentrated CBD cannabinoid extract suspended in a chewy chocolate caramel. The cannabinoid profile is primarily CBD with .03% THC.

In 1998 things took off again in England. Thanks to CBD's proven ability to degrade the psychotropic THC experience one gets from cannabis, GW Pharmaceuticals was able to convince the government they should be allowed to produce cannabis-based medicines that would not get people high. GW released their findings to the world showing CBD was more beneficial than originally thought and continued their efforts on the back of genetics provided by Californian-led Hortapharm B.V., the first farm licensed by the government of the Netherlands to grow research-grade cannabis. In 2015, GW's intravenous CBD was fast-tracked by the FDA for neonatal hypoxic-ischemic encephalopathy, which currently affects up to 24,000 newborns a year in the U.S. and Europe. In October of 2016, the U.K. officially declared the CBD molecule to be medicine. This dramatic announcement was spurred by the work of a nonprofit organization in California and followed in the wake of a mainstream television documentary that lead to CBD laws being enacted in conservative states across America.

In 2010, the organization Project CBD, lead by Martin Lee, helped bring the cannabinoid closer to the forefront of medicine by developing CBD-rich strains. This action marked the year CBD would become more widely available of California dispensary shelves. Since then, farmers have continued to push the medical envelope, now producing strains CBD to THC ratios as high as 30 to 1. These elite medical strains have found their way into a variety of delivery systems for medical cannabis patients including edible, topical and other concentrated options.

"Weed" was a special documentary hosted by CNN chief medical correspondent Sanjay Gupta that first aired in 2013. With this special, and subsequent follow-up documentaries, CBD hit the mainstream media in a major way offering many families a glimmer of hope for children battling severe seizure conditions. Today, many have been uprooted from across the U.S. in search of treatment in states like Colorado and California. These medical cannabis refugees have driven much of the progress seen in past few years. Sadly, the stories these families have on occasion been used by politicians to create empty medical cannabis laws, writing CBD-only legislation in states with zero access to medical cannabis, without providing a means for that oil to be produced, sold, or regulated. Despite these setbacks, the development of CBD-rich strains continues to grow and, as more and more research reveals the healing applications of this cannabinoid, we will surely see cannabis breeders creating additional strains showcasing the awesome power of medicinal cannabis.

CBD Garden [photo by Justin McIvor

Introduction to the Strains
by Ed Rosenthal

Breeding is not easy. It requires a keen eye, an acute sense of taste and most importantly, an ability to discern a plant with outstanding properties.

Breeders are a different kind of people than you or me. They are determined; they are, competitive, and driven by obsession. Breeders must be tough, tenacious and focused in order to pursue their careers. Not many people have this ability—a sort of perfect pitch in the area of THC and cannabis. In addition to skill, an inspired breeder has an intuitive ability to choose the right one. This book celebrates 49 right choices.

The following pages show some of the breeding done by seed companies around the world. All that follows are not just pretty pictures. These are the varieties and seeds that are part of the new industry. The strains are commercially available providing both home growers and licensed cultivation centers with predictable results.

portion of the book is divided into the following sections:
Indica: Strains that are 70% Indica or greater
Hybrids: Stains where the ratio is more equal
Sativa: Strains that are 70% Sativa or greater
CBD: Strains bred to contain a higher percentage of CBD than THC

Within each section the strains are listed alphabetically. There is also an index at the end of the book to assist with finding the page number of each strain.

From "seed to sale" it all starts here with the talent, patience and insight of the breeders.

[photo by Professor P]

Seed Companies & Varieties

True OG [photo by Elemental Wellness Center]

Indica

Indica plants originated around the 30th parallel in the Hindu Kush region of the Himalayan foothills. This includes the countries of Afghanistan, Pakistan, Tajikistan, Northern India, and Nepal. The weather there is quite variable from year to year. For this reason the populations in these regions have a varied gene pool. Even within a particular population there is a high degree of heterogeneity, which results in plants of the same variety having quite a bit of variability. This helps the population survive. No matter what the weather during a particular year, some plants will thrive and reproduce.

Pure Indica plants are short, usually under 5 feet (1.5 meters) tall. They are bushy with compact branching and short internodes and are arranged to capture all the light that falls inside the canopy. They range in shape from a rounded bush to a pine-like shape with a wide base. The leaves are short, very wide, and dark green when compared to most equatorial Sativas because they contain larger amounts of chlorophyll. Sometimes there is webbing between the leaflets. At the 30th latitude, the plants don't receive as much light as plants at or near the equator. By increasing the amount of chlorophyll, the cells use light more efficiently.

Indica buds are dense and tight. They form different shapes depending on variety. All of them are chunky or blocky. Sometimes they form continuous clusters along the stem, but their intense smells ranging from acrid, skunky, or musky to deep pungent aromas reminiscent of chocolate, coffee, earth, or hash. Indica smoke is dense, lung expanding, and cough inducing. The high is heavy, body-oriented, and lethargic.

Indica-Sativa hybrids naturally tend towards the Indica side of the family. They usually have controlled height. They don't grow very tall and after forcing flowering, their growth is limited. Side branching is usually not prominent so they can be grown in a small space. However, they have both Sativa and Indica influences, which may include surprising hints of Sativa in some aspects of the plant's makeup, flavors, or high. All strains in this section contain greater than 70% Indica and are listed in alphabetical order.

Plant Characteristics of Indicas

Height: 2' to 6' (0.6m to 1.8m)

Shape: Conical to bushy

Branching: Lots of side branching usually wider than its height

Nodes: Short stem length between leaves

Leaves: Wide short leaves, short wide blades

Color: Dark green to purple

Flower: Wide, dense, bulky

Odor: Pungent, sticky or fruity

High: Inertia, desensitizing

Flowering: 6 to 9 weeks

BARBARA BUD
House of the Great Gardener

70 I / 30 S • Even body/head, uplifting, giddy
• Citrus, berry

Barbara Bud is an Indica-dominant medicinal cross of a Shishkaberry mother and an Afghani Indica hybrid father that has extreme trichome coverage. The Shishkaberry mother is a favorite of the breeder, who has worked with compassion clubs across Canada for years. Barbara Bud, available since 2003, gets her name from a well-known Canadian radio host who was supportive of one of the House of the Great Gardener breeders during his constitutional court battle for medical cannabis.

GROW

Barbara Bud is well suited to indoor growing and adapts easily to size constraints in hydro or soil, staying small in small containers and going big in big ones. When vegetating, Barbara Bud branches considerably, forming an iconic Christmas-tree shape ornamented with crystal-frosted buds. During the 7 weeks of flowering, this strain doubles in size. At ripeness, buds are desirably compact — neither too tight nor too loose — and happily cloaked in trichomes. Some buds may take on a tinge of red when nutrients are properly flushed at the end of flowering. This durable strain shows good resilience to heat, cold, drought, and pests, making it a fine choice for beginners or commercial gardeners.

CONSUME

The citrus-berry smell hints at a terpene profile dominated by limonene yet balanced with others that produce floral, peppery, and herbal notes. THC levels that run upwards of 14% have made it a favorite in Canada's compassion clubs. Medicinally, it works broadly for depression, anxiety, inflammation, and acid reflux, while stimulating the immune system, inhibiting tumor growth, and reducing pain. The high comes on fast and lasts a long time, with an uplifting, creative, social effect that is more "up" than sleepy, making it nice for social activities that benefit from a bit of imagination.

Big Bang 2
Feminised Seeds

75 I / 25 S • Calm, stoney, long lasting • Smooth, fruity

Big Bang 2 is the offspring of two of the greatest: Northern Lights and Jack Herer. If cannabis originated from a primordial "big bang," these would be the progenitors of most modern strains. The classic Northern Lights, a tasty breeding Afghani/Indica, has figured in many Indica crosses. Jack Herer is a more complex variety, as befits the man who nearly single-handedly revived the hemp movement with his tireless activism and seminal book, *The Emperor Wears No Clothes*. His eponymous plant is an Indica/Sativa hybrid of some of the most significant strains, including Northern Lights, Skunk #1, and Haze. Big Bang 2 lives up to that lineage with big yields (up to 2.6 oz/ft^2 or 750 g/m^2) and high THC values (up to 22.5%) under optimal conditions. That's why Big Bang 2 has been Gorilla Seeds' top-selling seed for the past eight years.

GROW

Big Bang 2 generates big yields indoor or outdoor as a multi-branched plant. She starts off dark green with fat, Indica-style leaves and tight nodes, and has good tolerance to heat, cold, and pests. In flowering, the stretch will lead to medium heights with weight-bearing branches. She forms a sturdy pine-tree shape with some purpling at the end, orange hairs, and dense resin, but goes bushier if topped. Follow FIM methods for best results. Flowering is 8-9 weeks after a slow start. With enough light and moderate pruning to a multi-branch, the colas will form nice nuggets — very little popcorn bud. Production is impressive, with often a nice swell at the end, so don't get anxious and harvest before she's ready. Big Bang is what you get for your buck with this strain.

CONSUME

The flavor is smooth, sweet, and fruity with a touch of skunk — slightly pungent, but not cloying. The smoke is also smooth, with a long-lasting, relaxing, quality high. A good all-around smoke for hanging out with friends, watching movies, or "just chilling," it's also great for nighttime because of its restful, sedative effect. With a THC level this high, it is not the best choice for a boost at work, although it can be just the thing for stimulating appetite, lifting mood, or managing pain.

Big Buddha Cheese x Cheese Reversed

Big Buddha Seed Company

100% Indica • Uplifting, no ceiling, clear, long lasting • Old school, woodsy, fresh/pungent

The Big Buddha Cheese x Cheese is a remix of a classic strain. First available in 2004, it was reworked to Version 2 in 2014. The BBCxC mother is a coveted original UK Big Buddha cheese clone. The father is an original Big Buddha cheese reversed clone with a true reversed backcross of the U.K. The slender, elegant cheese leaves look a bit like a Sativa, but the Indica heritage is immediately evident in this strain's vigor, super-fast growth, and solid branching that spreads to soak up the light. Its electric lime-green color complements the compact foxtail buds that chunk up and get rich with calyxes near harvest.

GROW

BBCxC is a versatile plant suitable for both indoor and outdoor growing in whatever medium the gardener prefers. The Big Buddha Cheese x Cheese is a moderate to heavy feeder that loves all nutrients and boosters. This plant is a winner for all levels of expertise, but the odor of the plant is extremely strong and ramps up to super dank in the last 3 weeks, so stealth growers should have a plan for controlling telltale odors from reaching the wrong noses.

CONSUME

BBCxC lights up the senses with an amazing, slightly funky taste of sweet berry hash candy. Offering a high-grade, energetic stone with virtually no ceiling, BBCxC takes users on a thoughtful, long-lasting journey. The high is awake and cheerful, sometimes giggly and introspective, and always encourages an appreciation of the world. Some medical users say this variety has been effective for their chronic pain conditions, offering continuous relief. As uplifting as it is, it can also aid sleep, though it may leave you a bit thirsty, so have a beverage on your nightstand to soothe cottonmouth.

Bubba OG
Greenthumb Seeds of Canada

80 I / 20 S • Heady, eyedroop, sensual
• Fuel, coffee, floral, pungent

Bubba OG has a vaguely defined family tree, with a mother who likely hailed from the Hindu Kush and a ghost-cut OG Kush father. Bigger and heavier than either Bubba Kush or OG Kush, the Bubba OG has been selected and stabilized in a careful breeding program to create a strain with Bubba taste and OG potency. Canada's Dr. Greenthumb was the first to offer genuine Bubba Kush and OG Kush in seed form, and they have carried both plants in a continuous line, unchanged, for over 13 years. This happy marriage of those old-school favorites is sure to appeal to the Kush enthusiast.

GROW
Bubba OG has medium branching, which Dr. Greenthumb advises growers not to prune. This plant forms a basic cone shape with hard stick buds that become bulbous and dense with resin as they ripen. They can turn purple in the coldest climates, but are generally dark green with thick leaves. Bubba OG prefers more arid regions, withstanding heat, drought, and cold weather well, but struggles if conditions are overly humid, damp, or foggy. This plant needs about 9 weeks of flowering time, finishing in October at 45 ° North. While Bubba OG is not too challenging for a beginner, a discreet grower will take measures to contain her heavy, strong aroma in the garden.

CONSUME
The Bubba OG is a people's choice, with many fans who call it their favorite strain. Her smell is a divine mix of temple incense, flowers, and coffee that will entice even the occasional recreational user to take a nice long inhale. With its classic, fast-acting Indica effects and a long, powerful, yet happy stone, it goes well with sensual or satisfying chill activities and works well on pain.

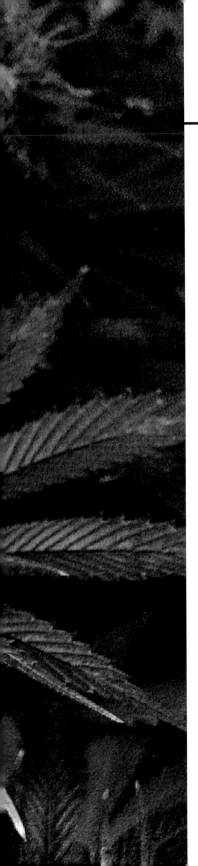

Cheesy Dick
Big Buddha Seed Company

95 I / 5 S • Knockout body • Strong, spicy

This nearly pure Indica with a touch of Sativa gets its cheeky name from a play on the popular British holiday treat, "spotted dick," a rolled pastry sprinkled with raisins, dried fruit, and creamy custard that's the UK version of fruitcake. When Big Buddha crossed a reversed father of his UK Cheese clone with a Spanish Moby Dick mother, the nasty name was irresistible — especially because the plant's taste is savory-sweet like the dessert. The variety's parentage includes some highly potent classics — Critical Mass, Big Bud, and White Widow — while Moby Dick has a reputation for rewarding growers with whale-sized yields foamy with resin. Mixed with what Big Buddha calls "Old Skool Classic," the result is a huge-yielding plant with classic flavors for the more serious grower.

GROW
Cheesy Dick shows her Indica nature in large medium-dark leaves and high yields. She does well as a multi-branched plant, producing hefty buds on widely spaced internodes and massive main colas. The plants perform best when topped. Odor-control steps will be necessary, as this is as dank a variety as we've come to expect from the Cheese family. Indoors, flowering time is 8-10 weeks. Outdoors, expect finish at the end of September.

CONSUME
One whiff of Cheesy Dick and the spicy cream cheese scent comes through with a hint of blackberry and fruit. The extremely musky dank of its aroma sends a message about this plant's extreme potency. A favorite of Indica connoisseurs, the flavor is equally strong, with the musky spice of Big Buddha Cheese dominating. The high resin content delivers a real knockout OG stone worthy of the heaviest Indicas.

Chemdawg

Greenthumb Seeds of Canada

90 I /10 S • Uplifting, playful, talkative • Dark cinnamon, hash, berry

In the opinion of veteran breeder Dr. Greenthumb, Chemdawg is one of the top-five all-star strains for its combination of effect, yield, taste, and ease of growth. He's not the only one who nominates this legendary strain for a place among top connoisseur picks. A gypsy strain that goes back decades, many rumors surround her genesis. The original Chemdawg is said to have come from the Dog strain that travelled with the Grateful Dead tours back when Jerry was still alive. Although parentage cannot be verified, reputation and rumor point to Kush and Sour Diesel. Chemdawg is both a breeder's strain, making up many successful crosses, and a "people's choice," getting excellent reviews for the happy, leisurely mood and medicinal value it delivers.

GROW

Chemdawg is a versatile indoor/outdoor strain with thick, upright branching, making it suitable to Sea of Green. Plants form a dense pine tree of serious, chunky, classically Indica buds. This strain is amenable to different gardening styles, and can work as well trained into multiple branches as it does in SOG gardens.

CONSUME

Some Chemdawg phenotypes may go more berry, while others may get dark and pungent with a diesel edge. The smell is strong and lingering. Her acrid taste carries a dark diesel edge, cinnamon hash and a hint of berry. The high is generally active and alert, calm, clear-headed, and cheerful, though the Indica body-high can be sedative and induce eyedroop. It can be a good bedtime strain for some but doesn't work for everyone. After the wave of sleepiness, many find it rebounds to a more energetic, creative mental state while still being relaxed. This makes it a good stress-management strain, but a refreshing beverage and some eye-drops are recommended to combat the cottony effects.

Cherry Cookies

Purple Caper Seeds

Indica dominant • Body stone, relaxed, couch lock • Cherry, gasoline

The Cherry Cookie strain is another winner from Northern California breeders, bringing out the best of the sweet Cherry Pie variety and a modern classic, Girl Scout Cookie. The Dark Heart Cherry Pie mother is a cross of Grand Daddy Purple from the U.S. and Durban Poison from South Africa, selected by Purple Caper as the best of three cuts. The father is a Girl Scout Cookie F2 backcross, consisting of Grand Daddy x Durban and OG Kush. The result is an Indica-dominant strain that Purple Caper calls a super-improved version of both parents.

GROW

Indoor cultivation is best for Cherry Cookies' Indica growth characteristics. This plant forms a symmetrical bush that averages 3 feet tall and 2 feet white. She prefers hydro and performs well with a multi-branch grow style, producing lots of branches with purple golf ball-size flowers. Expect dark, dense buds edging to purple with orange hairs. Purple Caper recommends keeping the environment at 80 °F (26.5 °C) during the light cycle and 60 °F (15.5 °C) when dark.

CONSUME

The aroma of Cherry Cookies is reminiscent of her Cherry Pie mother, with a pungent sweet-and-sour cherry smell that carries an acrid fuel note. The smooth smoke has a distinct cherry flavor, particularly on the exhale. As with her parents, this strain's intense effects are calming but heady, creating a deep physical relaxation that can tend to couchlock. Her appetite stimulation and nausea-fighting properties combined with her ability to fight pain and anxiety make this a popular choice for many medical users.

Cream Mandarine Auto® (SWS29)
Sweet Seeds

70 I / 27 S / 3 R • Mellow, sensual body high
• Fresh, fruity, floral

Released in 2012, this feminised auto-flowering hybrid boasts big yields, citrus aromas, and sweet flavors. Cream Mandarine Auto® (SWS29) was bred from an SWS06 Ice Cool elite clone mother—a NYC Diesel and Nafarroa Diesel with a global lineage of Spain-Holland-USA. Her father is SWS22 Cream Caramel Auto from an unusual, high-yielding Spanish genetic line with just enough Ruderalis to make her auto-flowering. Taken together, this hybrid is 70% Indica, 27% Sativa, 3% Ruderalis, with pronounced sweet mandarin citrus aromas from both her mother and father.

GROW

Cream Mandarine Auto® (SWS29) has a bushy structure and huge main colas of sticky buds. Thick leaves emerge from the main stem and first nodes of the side branches, but she has only a few thin leaves around the flowers. That makes this an easy-to-trim plant, though pruning should be avoided. She grows vigorously in hydro or soil right from the start, then explodes 2 weeks after germination, with flowering starting at 22-28 days without forcing, and a final height between 20 and 43 inches (50- 110 cm). A good quality soil mix suffices until mid-flowering, when low doses of liquid fertilizer can be added once a week. She produces well when grown in close spacing. Outdoors, between spring and the beginning of autumn is best. Finish is about 9 weeks after germination, regardless of latitude, thanks to the autoflowering genetics. Cream Mandarine Auto® (SWS29) can withstand a bit of cold, heat, or drought, but environmental extremes can negatively affect the final size of the plants. The yield is high for an auto-flowering plant.

CONSUME

This strain's smell is very discreet while growing and only slightly noticeable while flowering, though touching her flowers as she ripens releases the sweet mandarin orange smell the name suggests. The flavor is fresh, sweet, and slightly floral, with mature fruity tones, hints of earthiness, and an intense sweet mandarin background. The relaxing high is positive, optimistic, calming, mellow, and happy. Medical users report it helps with anxiety and sleeping disorders, as well as chronic pain.

Flowerbomb Kush

Strain Hunters Seed Bank

70 I/30 S • Sedative, giggly • Fresh, fruity, sweet

Named after a famous Dutch perfume, Strain Hunter's Flowerbomb Kush has a Green Crack mother with Skunk #1 mixed in. The father is an unknown Indica from Ohio crossed with a California OG Kush from an Afghan Hindu Kush landrace. This mostly Indica strain was selected out of a hybrid F1 as the most resinous individual, rich with cannabinoids and terpenes. Her color is dark green to brownish — purple, if nights are cold — with thick, very dark green leaves that turn to dark purple and gray at harvest time.

GROW

Flowerbomb Kush is a multi-branch plant that exhibits fast branchy growth, medium-long internodes, and usually a main cola at finish. Her slightly stretchy pine tree structure has long branches competing with main stem splits at the third or fourth internode. She does best with a short powder-feeding regimen and a flush in the last 7-10 days. Flowering is 8 weeks with peak resin production at week 7. This strain is good for indoor cultivation or outdoors in temperate climates. She withstands cold well, though she's not suitable for extreme outdoor conditions. Medium-hard to grow, Flowerbomb Kush is a vigorous variety in the right hands.

CONSUME

Flowerbomb Kush has a complex flavor profile that combines fruity-woodsy notes for a musky bubblegum candy taste. As with other Kush varieties, the high hits the body hard and fast with a sedative, couchlock effect, then creeps on the mind with giggly, long-lasting sedative qualities. This strain is good for relaxing, sleep, and pain control, but it is heady enough to create confusion, so it is best to avoid activities that require much focus.

GG#1
House of the Great Gardener

100% Indica • Even body-head • Sweet, hash, pine

GG#1, or Great Gardener #1, is a pure Afghani Indica. After many years of working with the original Afghani #1 clones, Mat, the breeder at House of the Great Gardener, created this seed strain based on it in 2011. Since then, GG#1 has become a bestseller in the Canadian compassion clubs that House of the Great Gardener supplies and has been used in their breeding program to create almost all of their medicinal strains (The Haoma, Digweed and Barbara Bud).

GROW

A pleasure to grow, GG#1 shows a strong structure. Bred for indoor cultivation in any type of growing method, this strain has great vegetative growth with lots of branching to a classic evergreen shape. Leaves are Indica-style thick, and the plant is short and bushy in true Afghani fashion. Plants double in size during vegetative growth but can be kept smaller by restricting container size. This strain produces high yields of conical, dark-green buds that change to yellow as she ripens. Flowering lasts 8 weeks. These are resilient, hardy plants with good resistance to suboptimal conditions and pests.

CONSUME

The fragrance of GG#1 is deep hash, floral, and smoky with woodsy flavors. The prominence of the terpene caryophyllene gives it a rich, spicy, almost peppery odor. Those terpenes, research has shown, have application to treat anxiety and depression. The strain is also notable for a dreamy, top-notch high that starts strong and delivers a relaxing body feel that is a mellowing pain killer. For those without particular health complaints, GG#1 offers welcome relaxation that pairs well with watching TV or a movie and winding down before bed. The effects are stony enough that it can be difficult to work or read with this strain.

M-42
Treehouse Genetics

90 I /10 S • Mellow, relaxing • Earthy, sour, blueberry

The M-42 strain from Treehouse Genetics is named for the Orion nebula, which shares the plant's pink and purple coloration. This predominantly Indica variety was bred from a Maine clone-only Blueberry Indica combined with the mostly Indica Cosmos Kush. The end product is M42, an extremely pungent Blueberry Kush that grows large, dense, colorful flowers thickly coated with resin glands.

GROW

M-42 generates prolific side branching to achieve a bushy spruce-tree shape. Treehouse recommends topping twice during vegetative growth to make a wide bush. Expect pink calyx growth in the interior of the flowers, regardless of temperature change. M-42 benefits from cultivation in soil with organic fertilizers to bring out her pungent flavors, whether grown indoors or out. Flush with clean water for 2 solid weeks pre-harvest for best results. After 50-55 days of flowering, this high-yield strain ripens near the end of September at 44 ° North latitude. Finish heights of 4-7 feet (1.2-2.1 m), with an average of 6 feet (2 m). A strong, pest resistant plant that is good for beginners and commercial propagation alike, she tolerates abuse and cold fall nights, no problem. Beware her quite strong smell while growing.

CONSUME

M-42's frosty, bulbous buds are light green in color, accented with pink and purple. The complex flavor of earthy sour blueberry carries floral-fuel undertones. The mellow high is very relaxing and a bit of a creeper but not a knockout. The relaxed, happy effect is peaceful, making it great for winding down after work. Medical users like M-42 for treating anxiety, stress, tension headaches, and body aches.

Peppermint Kush
Barney's Farm

90 I / 10 S • Heavy effect • Minty, peppery

Peppermint Kush lives up to its name with a spectacular minty aroma and taste, combined with the heavy-hitting high of a nearly pure Indica. Bred from crossing Stella Blue, a Sativa-dominant hybrid with a fragrant Blueberry background, and Nepali Kush, a cross of the classic Hindu Kush with a Nepalese landrace, this strongly Indica-dominant strain boasts THC levels upwards of 22% and registers more CBD than most strains this potent.

GROW

This strain can thrive indoors or out, though she prefers cool, temperate climes. Plants top out at under 3 feet (60-90 cm), making this a candidate for stealth gardening among the shrubs. Screen of Green (SCROG) is also an option, since she sprouts many side branches that fill out enough to need support in the final 2 weeks. Her large, dense colas can produce a beautiful array of colors, particularly with cooler temperatures as she finishes. This can be a very aromatic plant, so plan accordingly. Turnaround is quick and growth is strong, with flowering in as few as 50 days and yield up to a pound per square meter.

CONSUME

Peppermint Kush will blow your mind with an irresistible taste combination that starts with the peppery spice of a Stella Blue but leaves you with the minty aftertaste of Nepali Kush. On the toke, expect even more, as this strain's potency takes a back seat to none, but with a trippy Sativa-style edge that keeps your mind racing even if your body is locked to the couch.

Pliny Purple
Dusted Buds Genetics

80 I / 20 S • Creeper, even body-head, relaxing
• Wine, pine, fuel

The Pliny is a recessive GDP backcross, named for Pliny the Elder, the Roman who was the first to document the method of producing the deep purple color reserved for royalty. The idea was to backcross Granddaddy Purple parents, selecting to minimize the dominant grape-berry phenotype, until interesting genetic characteristics appeared. From germinating a large number of seeds, "the one" appeared that shared the dense bud structure of the GDP only with a musky aroma of Merlot wine. Since then, the breeders have further backcrossed those traits to stabilize them.

GROW

Pliny Purple responds to heavy feeding in hydro or soil and thrives indoors or out, as long as it's a dry climate. Node spacing is tight in early growth then spaces out with maturity. Strong side branches can support themselves well into mid-flowering before needing supports. This plant does well with mid-branches left on, so only thin the lower branches. Unpruned, Pliny Purple is medium bushy and spherical in shape, ranging from 36-60 inches (1-1.5 m). In flowering, color goes from light to dark green to purple, with a drop in temperature to help bring out the purple. Avoid overwatering during flush. This variety produces moderate to high yields of flowers so dense that good airflow and lower humidity are required. The very compact, evenly sized pine-cone shaped buds are dark green and purple with hairs that can appear very orange against the purple hues. Good for beginners, she is fairly resistant to pests. Mold risk can be mitigated by using a dehumidifier when curing. The yield makes her suitable for commercial uses.

CONSUME

Pliny Purple was selected for her unique aroma — a pungent, musky scent of Merlot wine tinged with pine and fuel, courtesy of her damascenone and myrcene terpene profile. Flavors are acrid with notes of cinnamon, lavender, nuts, and whiskey. The creeper high is long-lasting and evenly balanced between mind and body, calming the mind and producing a deeply narcotic physical relaxation.

Purple Caper
Purple Caper Seeds

Indica dominant • Dreamy, active, social • Tropical, fruity

The Indica-dominant strain Purple Caper was selected by the breeder as the winner from 50 crosses of some of the best Indicas in the western U.S. This fruity strain's mother is Grand Diesel, a cross of Grand Daddy Purple and Diesel from California. The father is White Indica from Las Vegas, a White Widow x pure Indica of unknown origin.

GROW

Purple Caper is recommended for indoor gardens, but plants can be grown outdoors in temperate climates such as California's San Francisco Bay Area or Central Valley. Plants show average branching for a mixed-parentage hybrid and will do well as a multi-branch plant in whatever medium you prefer. To bring out some purple color, adjust nighttime temperatures to dip under 60 °F (15 °C). Flowers on this plant get large, as do her crystals.

CONSUME

Both the flavor and aroma of Purple Caper hit notes on the sweet end of the fruity spectrum, reminiscent of tropical delights such as guava, mango, and papaya with a bit of grape. Have a smoke and the effect is active and blissful, with a creeper start and a dreamy, electric high end. This is a fine strain for playful pursuits, positive social scenes, or sensual fun.

Salmon River OG

Dynasty Genetics

80 I / 20 S • Blissful body, relaxation
- Sweet, berry, coffee, petrol

The Salmon River OG strain is named after the Salmon River in Oregon and grew out of the breeder's 2007 Huckleberry project through several backcross lines, as well as the Oregon Huckleberry IBL before its 2014 release. This Indica-dominant strain exhibits great stability and only three phenotype variations, all of which are "desirable keepers." The mother plant is a pre-1998 OG Bubba Kush from the West Coast. The male is a Blue Heron #111, which is a special combo of Blue Magoo x Blue Magoo/Huckleberry. Salmon River OG's Blue Heron father was selected from 120+ plants grown over two years, with Dynasty picking the plant that most resembled the original Blue Magoo. The recessive traits of the Blue Magoo mesh with the OG Bubba harmoniously, producing colorful, crystal-coated blossoms.

GROW

Salmon River OG is an easy strain to grow and requires very little maintenance for optimal results. The plant's exceptional bud-to-leaf ratio, medium height, and pole-like structure makes her easy to trim. Branching can be encouraged by topping at the third or fourth node and is best for non-SOG. This plant requires minimal nutrients, and Dynasty recommends beginning to flush the plants around week 5, due to their tendency to hoard nitrogen. Organic methods are recommended to bring out the best results. Stretch is about a foot (30 cm) after start of flowering; bloom times are quite fast. Plants average 3-5 feet (1-1.5 m). Salmon River OG resists heat, drought, and cold and is good with pests but can be prone to powder mildew outdoors or in humid or poorly ventilated spaces. Garden odors are mild until harvest time.

CONSUME

Salmon River OG buds boast a dense trichome coating with intense purple violet hues so dark some look black. Aromas of blueberry, coffee, cocoa, fuel, and lemon zest permeate, and most translate to flavor, especially when grown organically. The intense Indica high has an instant onset of head rush that reaches your toes and lasts 4-6 hours. This sleepy, relaxing strain has shown effectiveness for relieving nerve pain and anxiety as well as increasing appetite.

Strawberry Loud Cake

Loud Seeds

70 I /30 S • Awake, uplifting, creeper high • Fruity, minty, citrus

This Loud Seeds variety, originally Strawberry Loud Cake #46, is a cross of a female Headband and male Loud Cake (Girl Scout Cookies x Original Loud). This fast-finishing phenotype was selected (along with #12 that finishes even earlier) because it carries a rare recessive trait that results in pink pistils and a wonderful, unique aroma.

GROW

This strain works well as a multi-branch plant indoors or out, producing high yields of compact bulbous buds. A bright green plant with pink pistils that darken, Strawberry Loud Cake grows asymmetrically in a mostly upward shape — similar to sour diesel but without as much stretching. Prune while it is small. She responds well to any balanced fertilizer regime, but the nutrient regimen should drop at the end, and those new to gardening, this strain should watch the nutrient uptake. Finishes very early (end of September at 38° North), at heights between 9-13 feet (3-4.2m) outdoors. Indoors, plants that begin flowering at 2 feet end at 3.5 feet (1.1m). This variety will tolerate heat but not cold.

CONSUME

Aroma is the centerpiece of this variety, with the breeders at Loud Seeds selecting this uniquely fresh and minty phenotype from over 500 females. The flavor starts minty, too, followed by citrus and strawberries with a light musk note. The Strawberry Loud Cake high is airy and light, but also effective with an awake, Sativa-style vibe that creeps up. The psychoactive and medical effects can be enjoyed without getting heavy, allowing consumers to function through an air of dreaminess.

The True OG
Elemental Wellness Center

70 I / 30 S • Potent, analgesic • Pine, fuel

The outstanding characteristics of True OG have been recognized repeatedly as one of the top hybrid strains available. It has won many regional and national competitions and earned magazine covers, with a plant grown in Jamaica from Elemental seeds ranking among "The Strongest Strains on Earth" in 2015. The genetics of True OG are a closely guarded secret of Elemental Wellness, which has used this strain for more than a dozen years as the cornerstone for the development of other award winners, such as Mango Tango.

GROW

True OG is an Indica-dominant hybrid that retains some growth patterns of Sativas. Plants stretch with long internodal length and a low leaf ratio but produce large, dense buds. That makes for an easy trim, and expert growers can expect a large yield. Elemental recommends a Sea of Green garden, with plants that never get above 3.5-4 feet (1m) tall. Hydroponics and CO_2 supplementation in a sealed room will maximize yield. The maturation rate is 56-62 days.

CONSUME

The True OG aroma overflows with kerosene and lemony Pine-Sol with a hint of skunk in the background, all of which come through in the dank pine flavor. Testing by Steephill Lab shows that the True OG contains significant amounts of myrcene, humulene, and caryophyllene. The strain's dominant terpene, myrcene, has its own analgesic and anti-inflammatory properties and may help get THC and CBD across the blood-brain barrier, contributing to True OG's consistent, long-lasting effectiveness. With CBD levels that can exceed 2% and THC that can reach nearly 30%, the True OG makes powerful sedative and anti-inflammatory effects. This strain's stone is quick-acting and heavy but buoyant, with strong couchlock effects and intense cerebral euphoria. A popular choice for concentrate makers, True OG has been used to produce award-winning live resin.

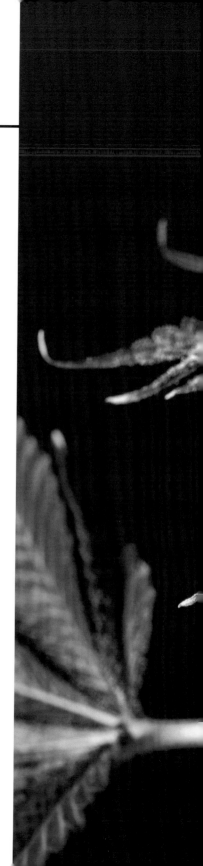

Wedding Cake
Dusted Bud Genetics

70 I / 30 S • Clear, happy, motivating
• Herbal, woodsy, hash

Wedding Cake is the 2011 offspring of a cherry phenotype mother of the AK-47 strain and an Indica-dominant LA Kush father. The dark-green fan leaves show its Indica leanings, and an inbred leaf curl/twist mutation marks this strain. The name mostly derives from the parents, Cherry AK47 and LA Kush, but the breeders felt it was no longer just a Los Angeles variety, so in naming it, sought to encompass the entire state.

GROW

Wedding Cake grows best in soil in a Sea of Green garden. She branches in a candelabra shape and can get asymmetrically top heavy, so cleaning lower branches is required, and supports are a good idea. Plants reach 5 feet (1.5m) max, with an average height of 42-48 inches (107-126cm). Early flower development is quick, with low to moderate yields. She produces well for a boutique strain, but is not suited to heavy commercial production or sloppy gardening. Wedding Cake needs modest nutrients with lower nitrogen to avoid stunting when young, and she burns easily if overfed phosphorus. That finicky nature means standard protocols do not apply. This is a strain for experienced gardeners looking for something distinctive.

CONSUME

While growing, Wedding Cake smells of sage than stone fruit — the cherry notes come later. The herbaceous aroma is not immediately identifiable as cannabis, making it easy to mask. The flavor is peppery with a woodsy clean draw of sage that reveals cherry undertones on exhale. Notes include hash, cinnamon, lavender, pepper, musk, sage, and sandalwood. The high is clean and clear and comes on fast but stays very lucid and functional. The effect is calming but not sedating. Wedding Cake can help relieve depression and anxiety, but is not as effective for pain as other Indica dominant varieties. This strain's high levels of the terpene beta-caryophyllene create a response similar to high-CBD strains because the terpene hits the same CB2 receptors.

Wubba

Animas Seed Company

75 I / 25 S • Stoney, munchie inducing, relaxation • Sweet, tropical, citrus

The newly released Wubba strain is a cross of a pre-1998 Bubba Kush mother that boasts Hindu Kush landrace genetics with an Icarus father (William's Wonder x Armagnac). This offering from Animas Seed Company — which gets its name from the Colorado river, Spanish explorers christened Rio de las Animas, or River of Souls — shares male parentage with their Sativa-dominant superstar, The Brotherhood, but takes its traits in the Indica direction.

GROW

Notable for her frosty floral structures, Wubba is a multi-branching plant bred for indoor cultivation in soil, but handles a wide range of climates just fine. The breeder recommends topping her once and feeding with a blend of synthetic and organic nutrients. Plants top out at 3 to 4.5 feet (1-1.5m) in height and typically starting flowering at a foot high and finishing at 3.5 feet. These dark beauties show heavy purples in the green, with tight, bulbous buds that look like frosty purple footballs. Even the dark green leaves can achieve a purple sheen. Wubba ripens in 8-9 weeks, though some phenotypes finish faster. This very hardy plant withstands climate changes well and bounces back from stress. A good choice for commercial propagation, Wubba does produce a strong smell in the garden that will need management if discretion is required.

CONSUME

Wubba has a nose full of grape, champagne, and sweet nectar. This fast-acting, long-lasting cerebral dome-rocker comes with a strong body stone that is relaxing to the point of being sleepy. Not surprisingly for a strain with a strong Indica Afghani background, Wubba has a powerful ability to combat anxiety, sleeplessness, and pain, as well as stimulate appetite. A great choice for decompressing on the couch after a long work day, this is not the best strain for mornings or mathematics.

[photo by Dutch Passion]

Hybrids

Cannabis originated in the Himalayan foothills and is now found all over the world. Until recently most of the genetic flow was one way. Seeds were carried to a new location. The plant adapted to its new environment in two ways. First, human pressure when growers selected the most vigorous plants that served their purpose. The second influence was the environment where plants adapted to latitude and climate.

This increase in diversity has been going on for thousands of years. Plants evolve, with each generation better adapted to its new homeland or growing conditions. Varieties look different, thrive under different conditions, and produce vastly different profiles of cannabinoids and terpenes.

It is only since the 1960s that breeders have been breeding the landraces to combine and modify their characteristics. The first wave of hybrids focused on manipulating cannabis to flourish indoors under lights. This was a counter measure to increased repression during pothibition. With legalization, hybridization has advanced; breeders now target for specific properties. As we have more scientific research about this plant and its components, we can create varieties to target specific medicinal, mental, and physiological effects.

Modern hybrids have complex parental histories. So many varieties have been crossed that it is a great mix-up resulting in many effective new combinations, so it is impossible to generalize about the qualities of their effects as well as taste and aroma and growth characteristics.

Indica-Sativa hybrids tend towards the Indica side of the blend in their size and structure. Their size and structure is a combination and compromise between the two groups leaning Indica. Their growth after forcing flowering is just a bit more than Indicas, which may grow only 10%. Indica dominant hybrids grow 15-30% taller and wider with narrow side-branches. This makes them excellent candidates for small space cultivation.

Sativa-Indica hybrids have more Sativa-like characteristics. They may grow 50% taller after forcing. Some S-I hybrids grow to double or triple their size when they are forced as small plants. Some varieties are harder to grow under lights and difficult in a sea of green without pruning because of the plants' space requirements. However, the Indica genetics do influence the plant size. The Indica genetics also results in earlier flowering and ripening, and the density, shape and size of buds.

Aside from striving for perfection by fiddling with the plants genetic variability, hybrids provide the cultivator with another advantage—Hybrid vigor. When plants are crossed the progeny has more vigor and often has higher yields than either of its parents. Ultimately, if you are philosophical you might ponder this: When you grow a hybrid, you are inheriting the wisdom of breeders who worked all over the world for thousands of years, combined with the passion of modern seed breeders who use their forbearers creation of landrace genetics as ingredients for their new alchemy.

American Beauty

Dr. Underground

50 I / 50 S • Calm body and mind • Blackberry, cherry soda, pineapple

When the famous TGA breeder Subcool decided to discontinue Plushberry, the breeders at Dr. Underground did not understand why. The strain was one of their favorites. They had a few packages of original Plushberry seeds acquired over previous years from shops in Spain, so they got down to work recreating it. Dubbed American Beauty, it carries the full Plushberry genetic heritage, including elements of Space Queen, Black Cherry Soda, Romulan, and the famous Cinderella 99. This delicate fruit in the form of cannabis is a perfect combination of Kush and some hints of pineapple from Cinderella 99.

GROW

American Beauty grows as you would expect of a 50/50 hybrid, branching easily, exhibiting beautiful shades of color, and popping out buds full of resin. Buds show some foxtails, and you'll notice that vegetative leaves bow up initially then return to normal in flowering. Big containers of soil generate the best results outdoors, while aeroponics is the recommended method indoors. This uniquely aromatic strain presents as two similar phenotypes — the first showing the reds and purples of Black Cherry Soda, the second having a bit less color but larger, denser buds. In both cases, plants grow very vigorously, resist mold, and produce a moderate yield.

CONSUME

American Beauty does not reveal her special berry-drenched skunkiness until you break open the bud. Once you do and fire it up, you get the full, super-tasty flavor spectrum of blackberry jam and cherry soda with a touch of pineapple sour. On the stone, the Indica dominates, relaxing body and mind in equal measure. This is an excellent strain for managing stress.

The Brotherhood

Animas Seed Company

60 S / 40 I • Alert, electric, trippy • Chocolate, woodsy, spice

The Brotherhood is the Animas Seed Company's stunningly strong Sativa-leaning hybrid, the offspring of a Slick Willy mother from Colorado crossed with an Icarus male of Afghani heritage by way of Colorado. The Slick Willy mother is herself a cross of a Bubba Kush/Chem 91 x The White/Chem 91, while the Icarus father marries William's Wonder and Armagnac. This pungent and nearly hallucinogenic strain was developed for, and named in honor of, the breeder's good friends, The Chris Robinson Brotherhood, the psychedelic rock 'n' roll band from California.

GROW

Ideal for indoor gardeners seeking top-potency genetics, this multi-branch strain grows squat and bushy despite its Sativa dominance and boasts an incredible crystal content. These plants appreciate Screen of Green training and thrive with a blend of synthetic and organic nutrients in generous quantity. Flowering takes 9 weeks on average, finishing at an unpruned height of between 3-4.5 feet (1-1.5m). No special ripening techniques are needed, but colder temperatures at finish bring out the purple. The bulbous, frosty flowers are so white with trichomes they look like they have been rolled in sugar, with purple hues peeking through. Even the thick, dark-green foliage show early trichome development… all the way to the tips of the fan leaves. This fairly hardy plant adapts well to varying conditions, handling cold, heat, and drought, and some phenotypes resist powdery mildew. Her forgiving nature makes this both a good beginner's strain and a sound choice for commercial cultivation.

CONSUME

On the nose, The Brotherhood has a clean, earthy aroma with fuel undertones. The creamy flavor is nearly doughy with a nice balance of spicy and sweet that has chocolate and coffee notes. One of the strongest strains available, testing at 34% total cannabinoids and more than 29% THC, The Brotherhood nonetheless imparts an active, alert high that stays surprisingly functional. The trippy electric energy is modulated with a blissful undercurrent that also carries a bit of eye droop.

Frisian Dew

Dutch Passion

50 I / 50 S • Balanced, cerebral, peaceful • Earthy, spicy, citrus

Available since 2007, Frisian Dew is the high-yield offspring of a potent, easy-to-grow Super Skunk mom that came out of the Dutch Passion Skunk breeding program of the 1980s. Her Purple Star father was selected from a purple strain noted for hardy, robust growing in outdoor conditions. The award-winning Frisian Dew took several years of clone selection and selective breeding to stabilize, and is now Dutch Passion's most popular outdoor variety. Carefully selected for outdoor conditions in Holland, the name stems from an ancient Germanic ethnic group native to the coastal parts of the country. Frisian Dew is a prolific producer of delicately colored, pink and light-purple buds with a density neither too tight nor too fluffy. Her large, thick leaves turn purple in some growing conditions, and buds on outdoor plants can be quite large.

GROW

Frisian Dew has been bred with a tolerance for outdoor conditions, but she can be grown indoors, too, of course. This branching plant will take all the space allowed, starting with a broad-base until it starts to stretch. Unpruned, Frisian Dew looks quite gangly with a main bloom that will reach great heights. Don't bother pruning outdoors — leave her to her own devices and watch her go. The pistils are often pink, as are the trichomes in some plants. This strain needs amply nitrogen in vegetative growth to maximize height — 9 feet (3m) is an average, while 12 feet (4m) is possible in the outdoor sun. Flowers are ready after about 8 weeks of flowering, or start of October outdoors. Yields are bountiful, with some growers reporting up to 5-6.5 pounds (2-3kg).

CONSUME

The aroma of Frisian Dew is mild to medium, while the flavor is earthy spice with hints of citrus and fuel. She generates a strong, fast-acting, quality high that is balanced and builds quickly. The feeling is peaceful and relaxed with a floating, detached, anesthetic effect. Strongly cerebral, this strain is not the best for socializing with crowds or groups that include straights, but it is a good choice for movies and simple pleasures.

GSC
Greenthumb Seeds of Canada

60 S / 40 I • Active, eyedroop, and physical • Coffee, hash, fuel

Dr. Greenthumb's GSC, or Girl Scout Cookies, variety was sourced for them by cannabis legend Jim "Dogless" Ortega. GSC was bred from an OG Kush mother that is a Chemdawg x Hindu Kush cross from California, while the Cherry Pie father is a cross of Durban Poison from Africa and Grand Daddy Purple from the U.S.

GROW

This strain grows enthusiastically indoors in hydro or soil. Sea of Green technique can be used, but plants grow fairly tall and generate lots of branching. This strain can produce amazing trichome coverage on sun leaves. Plants respond well to all the typical ripening techniques, so as dark period increases with some phosphorus and calcium and a temperature reduction boost the blooms. Plants can turn purple and start looking like a shorter Indica plant but stretches a bit. Finish time is late October.

CONSUME

GSC generates a connoisseur-quality product, with the predominant flavors and aromas leaning to fuel and hash. The trichome coverage makes it a good candidate for hash making. On the toke, the high is long lasting, active, and alert, with a cheerful, happy effect that can get giddy and giggly. The Indica side comes through in a physical wave that can leave you with eyedroop. GSC is a good strain for enjoying with coffee and friends.

[photo by Gracie Malley]

Loud Scout
Loud Seeds

50 I/ 50 S • Even body-head • Licorice berry pine

Loud Scout turns up the volume on a classic strain. This Northern California strain captures the desired traits of the original Girl Scout Cookies with an improved yield and unique flavor notes in a strain that tops the scale for potency. Loud Seeds created Loud Scout by crossing a GSC mother with a Platinum OG Kush father, and selected the phenotype based on yield and aroma.

GROW

Loud Scout is a bright to lime-green plant with moderate node spacing, purple tint in flowering, and compact, bulbous colas like swollen pine cones. She grows well as multi-branch or using screen of green in any medium, indoor or outdoor. Plants have an upward, asymmetrical shape that requires regular de-leafing of the undercarriage to achieve optimal flower size and structure. Be alert for some hermaphroditic tendencies, which can be mitigated by providing ideal conditions and climate. Drop the nutrient regimen slowly. This plant can reach 8 feet (2.4m) when unpruned outdoors but will cooperate at smaller sizes when grown indoors. Yields are moderate.

CONSUME

Loud Scout smells of licorice, berry, pine, and musky cherry. Flavors are earthy and creamy with black cherry tea notes. On the smoke, effects are quite strong and long lasting with a bit of couch-lock, increased appetite, and good pain relief that medical users appreciate. Body relaxation and heady, narcotic intensity dominate. Recovering from hard activity and sleeping are good activities for this strain, courtesy of the 28% THC and 2% CBD in the cannabinoid profile.

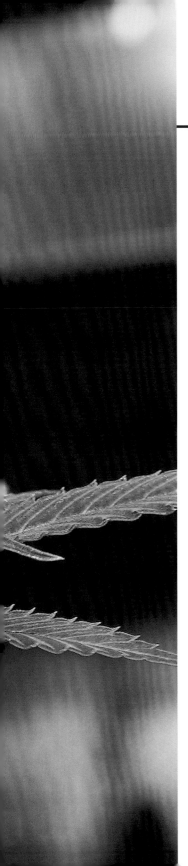

Miami Blues
Loompa Farms

60 I / 40 S • Alert, cheerful, body high
• Pungent, sour, blueberry

To create Miami Blues, breeder Jollymon crossed a Miami White mother (White Widow x OG Kush from Florida), with DJ Short's Blueberry from Oregon (Oaxacan Gold x Chocolate Thai x Highland Thai x Afghani). The Miami White mother is a beautiful girl to behold but lacked an enticing aroma. Adding Blueberry genetics lent a distinctly sweet smell and calmer effect. Over 200 seed starts were grown and observed in the breeding process, with the most desirable selected for the release of this variety.

GROW
Miami Blues performs best indoors with light-deprivation grow styles in any medium. The slow vegetative growth of this plant leads to smaller plants with lots of branching. Unpruned plants will stay bushy and branch excessively, so Loompa recommends pruning to thin and space. Sea of Green will also work. Lowering the grow room temperature by 5-10 °F (2.5-5 °C) during final flush reduces the stress from lack of nutrients.

CONSUME
The dense, sticky buds of Miami Blues emit strong aromas of sweet blueberry with a fuel undertone that surfaces after the blueberry fades. Taste is sweet and sour with a pungent, fruity, fuel flavor that has a candy berry finish. The long-lasting high of Miami Blues is alert and ready for action, cheerful and clear with a blissful, body-relaxing effect. Happily euphoric, this strain is well suited to social fun of all sorts.

Strawberry Banana

Crockett Family Farms

---

60 S / 40 I • Dreamy, heady, relaxing • Strawberry, hash

Released in 2012, Strawberry Banana is the award-winning offspring of a Banana OG mother (Banana Kush x OG Kush) from northern California crossed with Serious Seeds' Bubble Gum strain from Amsterdam. The fruity result is an extremely resinous Sativa-dominant hybrid that emits a strong strawberry smell in the garden.

GROW

These plants thrive indoors when grown in soil with organic fertilizers. Strawberry Banana's classic conical shape can reach up to 14 feet (4m) requires some management, with the breeder recommending multiple toppings — though not the banana-split kind. Indoors expect 6-9 feet of height and medium to high yields. Her thick, large leaves feed resinous, spear-shaped, dark-green buds featuring bright red hairs. As with many varieties, flushing for 10-14 days pre-harvest improves the final product. Flowering averages 60-65 days indoors, or between October 15-30 outdoors in California. Her vigor, ease of cultivation, and resistance to rot makes this a good choice for anyone from beginners to commercial gardeners. Plants handle cold and heat fluctuations fine, so long as they do not get too dry.

CONSUME

Strawberry Banana lives up to its name with an assertive aroma of sweet strawberry fruit and a flavor that combines strawberries and hash. The long-lasting high features a functional, inspired effect that comes on fast and strong then tapers slowly. The strain accentuates euphoria, happiness, and contented feelings, complementing nearly all outdoor activities, art, and moments of relaxation. You may get some munchies and eyedroop, and it is not the ideal strain for focusing or difficult physical tasks.

Sweet Tarts

Unknown Prophet

60 I / 40 S • Euphoric, relaxing • Pungent, citrus, pine

Sweet Tarts is a balanced Indica-dominant cross of a Derbs mother from South Africa with an Orange Tahoe father from NorCal. The Derbs was a selectedDurban Poison S1, while the Orange Tahoe was the winning selection from an old-school Cali-Orange crossed with Tahoe OG. The goal was to blend an exotic landrace variety with some new favorites from the Orange and Kush families.

GROW

Sweet Tarts thrives indoors or out, with a predilection for the outdoor climates of Northern California and the Pacific Northwest. These exceptionally vigorous plants do best in soil, branching enthusiastically, though her speed from clone to harvest makes her well suited to hydro and Sea of Green cultivation. Trellis netting is the breeder's recommended method. Left unpruned, plants grow out to a classic cone shape and can exceed 15 feet (5m) in height, with 5-8 feet on outdoors and 1-3 feet indoors. She responds well to topping, super-cropping, and defoliating, so don't be shy with the clippers. The dark hunter green color of the thick leaves fades in flowering to a light lime, but then at finish the darkness returns, with some phenotypes literally fading to black. Her compact, bulbous buds range in color from light lime to black-purple, but show high-contrast bright orange hairs amidst the trichomes. Calyxes can look like they are eating each other, and tops can fill out to the size of 16 oz water bottles. Lowering temps can coax out even more color, but don't go below 65 °F (18 °C). Less is more with nutrients, and don't rush your flush – water only for the last 2-3 weeks. She prefers heat but tolerates cold and resists pests and disease. Plants typically finish in 56-63 days indoors or mid-September at 39 ° North, but some phenos will ripen in as few as 49 days The astringent garden aroma can require management, but Sweet Tarts' forgiving nature makes this is an excellent choice for beginners. Yields can be very high with the proper technique.

CONSUME

The acrid aroma is so powerful you can almost taste the terpenes, though the character of the smell differs from phenotype to phenotype, with notes of Orange and Kush. What comes through on the palate is a flavor that is both sweet and tart. The Indica-Sativa balance relaxes body and mind though the Indica effects dominate as dosing increases. Generally, a clear, happy high that leans to the euphoric, making it great for social events.

White Widow

Dutch Passion

65 I / 35 S • Even body-head, euphoria • Floral, fuel

An internationally renown strain, White Widow is a 1990s classic from the masters at Dutch Passion. Created by crossing a Jungle Brazilian Sativa mother and a hybrid father containing south Indian, this strain is a staple of Dutch coffee shops and seed banks. Her potency and appearance set her apart from the crowd, with a name reflecting both the white frosting of resin that covers plants and her hard-hitting effect. The original buds were rated as "exceptionally potent" by the Dutch Passion test-team, a group with predictably high tolerances, and their judgment has been borne out by repeated award wins.

GROW

While slow to gain height and structure, White Widow's growth in flowering is so vigorous plants occasionally grow buds on leaves. The side branches surrounding her massive main cola can become quite heavy with sufficient light. Dutch Passion recommends removing lower branches to allow the plant to focus her energy. Medium-strength fertilizers are fine, but this variety can take heavy feeding in the last month. Light green leaves often go darker as resin loads up during flowering. Two days of total darkness before harvest maximizes the frosting. The light emerald green of her bulbous, oval buds contrasts with burnt-red hairs and a silvery sheen of resin for a look that shouts out, "vaporize me." A forgiving indoor/outdoor variety that can withstand cold, heat, and drought, White Widow shows her best side under skilled hands in optimum conditions. The famously thick resin protects her from pests, though the discreet gardener will need to guard against her unmistakable stench. Most harvest after 9 weeks of a 12/12 light regimen.

CONSUME

Once cured, White Widow's garden stench translates to a pungent yet pleasant aroma and rich, fuel-themed flavors with hints of coffee and sometimes garlic. The hard-hitting high of White Widow is cerebral, with plenty of euphoria at the start. An "awake and up" type of buzz in smaller doses, sleepiness dominates as you get more onboard, especially 2-3 hours after onset. Delirious, blissful, detached effects enhance activities such as walking and listening music or visiting with friends. This cerebrally active strain is too much for reading books or watching complicated movies, but White Widow has good all-round medical properties.

Yeti OG
Loompa Farms

60 I / 40 S • Even body-head • Fresh, fruity, floral

Yeti gets its name for its size and the geographical location from which it sprang in 2008. Bred in Bigfoot country, this Loompa Farms strain is a cross of an Underdog mother and a Black Domina x Snow father from Humboldt County. Dad was short and wide and a heavy yielder, while mom was crystal-coated and awesome.

GROW

A lanky but dense plant with thin leaves, Yeti is very OG, producing buds that are long, fat, and quite compact. Yeti was bred to be a tall, branching outdoor strain, but she can also be grown indoors. Yeti grows in a mostly up pattern that needs good support for its many branches and pruning of the bottom ones. Loompa Farms recommends soil for this variety, but this strain is a glutton for feeding and puts on weight when you give her what she wants. Ripens in October in Humboldt. To ripen indoors, increase the dark period and change the temperature. Resists bud rot and loves heat. It is not difficult to grow, so it is good for a beginner or commercial grower, so long as the strong smell can be tamed or ignored.

CONSUME

This strain has an earthy citrus aroma and a creamy, rich smoke with a smooth lemony flavor many connoisseurs rate top shelf. The strong Yeti high contains a nice combination of up and down effects, grabbing you on the first hit and holding on with a long-lasting, narcotic stone. Use caution if doing anything that requires concentration or has consequences if you can't stay focused.

[photo by Nadim Sabella]

Lands End [by Norstar Genetics]

Sativa

Sativa plants grow from the equator through the 50th parallel. They include both marijuana and hemp varieties. The plants that interest marijuana growers come from the equator to the 20th parallel. Countries from this area are noted for high-grade marijuana and include Colombia, Jamaica, Nigeria, Congo, Thailand and Sumatra. Populations of plants from most of these areas are quite uniform for several reasons. Cannabis is not native to these areas. It was imported to grow hemp crops and then it adapted over many generations with human intervention. Each population originated from a small amount of fairly uniform seed from the 45–50th parallel. Then the populations evolved over hundreds of generations with the help of humans. This led to fairly uniform populations in climates that varied little year to year.

Sativas grow into 5–15 feet (1.5–4.5 meters) tall symmetrical pine-shaped plants. The spaces between the leaves on the stem, the internodes, are longer on Sativas than Indicas. This helps to give Sativas a taller stature. The lowest branches are the widest, spreading 1½ to 3 feet (.5–1 meter); since the branches grow opposite each other, plant diameter may reach 6 feet (1.8 meters). The leaves are long, slender, and finger-like. The plants are light green since they contain less chlorophyll.

Sativa buds are lighter than Indicas. Some varieties grow buds along the entire branch, developing a thin but dense cola. Others grow large formations of more loose, spongy buds. The smoke is sweet and spicy or fruity. The highs are described as soaring, psychedelic, thoughtful, and spacy.

Sativa-Indica hybrids tend towards the Sativa parentage. They are taller plants, which will grow to double or triple their size if they are forced when they are small. They are usually hard to grow in a sea of green, as the plants demand more space to spread out. However, the Indica genetics may influence the size of the plant or its buds, the speed of flowering, the density of buds, the flavors, or the high. All strains in this section contain greater than 70% Sativa and are listed in alphabetical order.

Plant Characteristics of Sativas

Height: 5' to 25' (1.5m to 7.5m)

Shape: Tall, Christmas-tree shape

Branching: Moderate branching, wide at its base, single stem at top

Nodes: Long stem length between leaves

Leaves: Long leaves, thin long blades

Color: Pale to medium green

Flowers: Long sausage-shaped flowers

Odor: Sweet to spicy

High: Psychedelic

Flowering: 8 to 15 weeks

Arturo

Evolve Marijuana

70 S / 30 I • Energetic, clear, munchie inducing • Earthy, musky, pine

Arturo is a Sativa-dominant hybrid that makes the most of old-school genetics. The mother is a clone-only, pre-1989 Alaskan Thunderfuck whose Sativa genetics are something of a mystery. The poppa plant was a pre-1998 Kali Mist from Serious Seeds in Amsterdam. This new strain created in 2005 by the Cannabis Seed Co., a licensed producer in Washington State, is sold under the brand Evolve Marijuana.

GROW

Arturo grows great indoors and outdoors, flourishing in both hydro and soil, but prefers all-natural nutrients and a biological program that allows her to feed as much as she wants. This versatile strain shows strength and determination wherever she is cultivated, even outdoors in Northwest Washington State, where it was developed. A squat but sturdy sort, Arturo boasts beefy branches to handle large, crystal-laden flowers. Plants like to go bushy, but with minimal pruning she can reach 10 feet (3m) outdoors. The breeder recommends a top and pinch approach with low-stress training. Topped plants generate a canopy like an oak tree, full and rounded. In a controlled setting, Arturo enjoys 13 hours of darkness and 11 hours of full-on LED light. As she starts to flower, her compact buds look like late-stage dandelions as they fill with pink hairs and sparkling crystals. Her yam-shaped buds can resemble the Thai sticks of old, with nugs flowing into each other. This is an extremely hardy variety with a strong smell that resists pests and disease and bounces back from almost anything. Yields are reliably large. Currently being grown commercially, Arturo is also a great strain for the novice cannabis cultivator.

CONSUME

Arturo has a very earthy, musky, pine aroma with pungent Haze qualities that proceeds smoothly from nose to taste buds. The amazing high hits almost instantly and climbs steadily to an enjoyable plateau that lasts for many hours before making a very mellow descent.

This uplifting strain inspires creativity and is great for pain. While nice to share with a friend over interesting conversations, the complete experience is felt more on a personal level and can be wonderful to just experience all by yourself. Put on some headphones and literally feel the music. Write and feel what you hear so clearly in your mind flowing on the page. Walk and give deep thought to your future. Arturo is good for anything that benefits from having a focused creative mind.

Caboose

Strain Hunters Seed Bank

70 S / 30 I • Trippy, stony, intense • Fruity, metallic

Caboose starts with a Trainwreck mother, the popular super-Sativa strain from the "Emerald Triangle" of Northern California that is widely sought for its soaring, lucid high. Strain Hunters crossed that classic strain with a Salmon Creek Big Bud father, a mostly-Indica Big Bud phenotype that is also from the U.S. The result is a delicious-smelling strain with impressive yields. The famous California elite cut of this strain was reversed by Strain Hunters and released in a feminised form in 2012.

GROW

Whether grown in hydro or soil, Caboose quickly becomes a densely-branched plant with small internodes. The breeder recommends leaving it unpruned, but it can also be reduced to several little branches for a Sea of Green. This short, bright-green plant forms thick leaves and spectacularly chromatic colas with photo-worthy color. At maturity, the buds become pear-shaped, finishing about 8 or 9 weeks after forcing. Caboose likes powder feeding, but be sure to flush for the last 7-10 days at the end of flowering. This strain dislikes unseasonable weather but resists hot conditions well. A medium to hard strain to grow, it is best as a challenge for intermediate gardeners and not recommended for beginners.

CONSUME

Caboose has intense fruit flavors that start quite sweet but deliver a strong copper aftertaste. A down high with trippy insights, Caboose is cerebral at first, delivering an enjoyable intellectual or philosophical mood, then melts the body with a sedative, body-stoned feeling. Relaxing activities including reading will be doable, but nothing too active. The intense, calm high produces deep body relaxation with a trippy dimension and strong appetite-boosting effect.

Cherrygasm

TGA Genetics

---

70 S / 30 I • Stimulates the mind and appetite • Dark, floral, musk

Just released in early 2016, Cherrygasm is the sweet child of a clone-only Cherry Pie OG mother from Northern California crossed with a Space Queen father from Canada to produce more cherry flavor in a seed-friendly plant. Mom comes from crossing two classics: Grand Daddy Purple and Durban Poison, while dad is a cross of Romulan and Cinderella 99. The variety that results is a Sativa-dominant hybrid that will satisfy your sweet tooth.

GROW

Cherrygasm is a multibranch variety that averages 4-5 feet (1.5m) and prefers soil, whether grown indoor or outdoor, and organic nutrients. Top early to get four-headed shrub-shaped plants, or let it go unpruned for a small spruce shape that can reach 6 feet (2m). Flowering is 8 weeks, or mid-October outdoors, with colors shining through at 6 weeks. Cooler night temperatures will help Cherrygasm show her true colors, morphing from a bright lime green to a rainbow at harvest. Her compact, triangular buds run from purple to red through a nice tinge of magenta, finishing so thick with resin hash makers flock to her. The rock hard buds make her more resistant to mildew and mold, and feedback from growers Indicates this is a robust strain that stands up to stress. Optimum conditions are 78 °F with high humidity in a dome, but this plant is very easy to cultivate and process, with moderate yields.

CONSUME

Cherrygasm reeks of cherries, lemons, and diesel with musty lavender notes. Her flavors are similar, with an added smidge of Skittles. The high has immediate onset and lasts a good 3 hours. Despite its Sativa dominance, the heady smoke is very calming and relaxing. The stony stimulation hits your mind and appetite hard. While not recommended for sports or driving, Cherrygasm is conducive to creative pursuits such as writing or making art.

Eagle Bill
Sensi Seeds

70 S / 30 I • Energetic, social, uplifting • Piney, citrus, pungent

Launched at the beginning of 2016, Eagle Bill is the astonishing product of a secret combination of genetics, including Pure Haze, Colombian, Thai and Jamaican. Created as a tribute to Eagle Bill Amato, the Cherokee medicine man who popularized vaporizing; this strain has a Haze backcross mother and a hybrid Thai and Jamaican father. One of Eagle Bill's last wishes was that he be a part of the breeding process of his namesake strain. Sadly, he was unable to experience this uplifting hybrid with its pungent, pine-forest aroma and richly satisfying flavor, but everyone involved at Sensi Seeds is confident that he would have thoroughly enjoyed the culmination of this very special project.

GROW

This strain favors indoor environments in either soil or hydro but will grow well outdoors in warm southern latitudes. As a multi-branching plant, it is well suited to SCROG techniques or super-cropping. Eagle Bill exhibits so much lateral branching an unpruned plant will resemble a candelabra. Compact, bulbous, light-green buds form equally compact groups of calyxes amid thin, light green leaves. Wide internodal gaps and the spiral of buds around the stem and branches show its Sativa heritage, though buds are dense and heavy.

This variety has shown different phenotypes represented in its lineage, with two being prominent. One is a very Sativa-like phenotype, developing large buds covered in fox tails. The other is shorter and more skunk-like, with a more compact bud structure. Being a Sativa-dominant variety, this plant can prove tricky for first-time growers. Moderately smelly in the garden, flowering runs 60-70 days. Plants produce medium to high yields.

CONSUME

Eagle Bill has a balanced pungent pine and citrus aroma and flavor that is as ideal for vaporizing as you'd expect from a strain named for the King of Vape. The high is energizing and uplifting, leaving you awake, happy, and giggly. While not great for deep relaxation, this is a good choice for social interactions or creative pursuits.

Good Dog

Greenthumb Seeds of Canada

70 S / 30 I • Uplifting, heady, happy • Hashy

Good Dog is a cross of a Chemdawg mother and a Mexican strain called Queso Negro, both of unknown origin. The heritage of Good Dog's parents may be unknown, but Dr. Greenthumb's breeders were certain both were fabulous strains that would yield a special plant if crossed. They were right.

GROW

These plants do well indoor or outdoors, growing wide like pine trees with big colas and little branching. That makes them well suited to Sea of Green in hydro or soil. Dr. Greenthumb recommends increasing the dark period and changing temperature and fertilizer at ripening time. Harvest time outdoors is October.

CONSUME

Good Dog has a fresh citrus and floral aroma overlaying a pungent hash odor and strong hashy taste. On the toke, this strain hits an active, alert high that is clear and lucid with a cheerfulness that can be giggly and playful, making it great for happy social occasions. Its uplifting effect can take you soaring, but expect a bit of eyedroop, too.

Grandaddy Purple Platinum Cookies

Johnston's Genetics

70 S / 30 I • Even body-head • Musky, grape, fuel

Casting a Candyland mother to play opposite a Black Sugar Berry father may sound like the makings of a 1970s blaxploitation movie, but the colorful result is contemporary. The breeders at Johnston's Genetics put them together because they knew the aesthetics and flavors of the two sweet strains complement each other and improve yield characteristics. The Candyland mother is a combo of Bay Platinum Cookies and the famed Grand Daddy Purple from California. The Black Sugar Berry father is a Sativa-dominant hybrid, mixing Sugar Black Rose with Plushberry from Colorado. About two-thirds of GPPC plants will turn purple with leaves that are golden and purple. Her bulbous purple buds covered in trichomes form fat teardrops with spots of green. Calyxes range from magenta to green to purple, and pistils are vibrant orange and gold.

GROW

Granddaddy Purple Platinum Cookies is a cooperative strain with lots of branching that will thrive indoors with almost any method. Outdoors, she is good for cooler, temperate climates and mountainous regions where seasons are short. In the slow-growing vegetative stage, plants are a dark, luscious green and benefit from going slow to add yield-boosting structure. Halfway through 8-9 weeks of flowering, she starts turning to dark purple, even without cooler temps, but the effect is more pronounced when the thermometer drops, with leaves becoming gold, purple, red, and even magenta. Going dark for the 48 hours before harvest also brings out the color and trichomes in these short, robust plants. Just keep it above 55 °F (12.5 °C). A good strain for beginners, GPPC reserves her maximum potential for the experienced grower.

CONSUME

The aroma of Granddaddy Purple Platinum Cookies conveys grape and musky fruit with undertones of fuel. The smoke carries a smooth, earthy flavor of sage and sandalwood with a hint of berry. The high-quality stone is awake but relaxing, with a smooth start that creeps quickly to a strong, calming, clear-headed high. A good choice for meditation, fishing, light reading, leisurely hobbies, and social events, it also treats anxiety and PTSD well. GPPC is not recommended for marathon running, sword juggling, operating heavy machinery, or flying airplanes.

Grandma's Sugar Cookies
Johnston's Genetics

75 S / 25 I • Energetic, social • Floral, mint, fuel

Grandma's Sugar Cookies is a Sativa-dominant hybrid that builds on great strains from California and Colorado. The Girl Scout Cookies mother is a California cross of an OG Kush x F1 Durban, while the Black Sugar Berry father combines Sugar Black Rose x Plushberry from Colorado. The Girl Scout Cookies strain is a Johnston favorite for its growth structure, trichome content, and outstanding smell and flavor. Add the strength, yield, and flavor of the Black Sugar Berry, and the breeders knew they would have a winner. This strain is dedicated to the beloved LeDona Call, who over a 97-year life became renowned for the sugar cookies she baked for her more than 250 foster children.

GROW

Grandma's Sugar Cookies shows vigorous Sativa-like growth patterns with considerable branching, especially if topped, and heavy yields of colorful bud. Plants develop as wide cones with a large middle cola and thick, sturdy side branches. This strain should receive only minimal pruning in vegetative growth, but topping will bump up the yield. Thin, Sativa-structured leaves are vibrant green at early flowering. As flowers plump up, both buds and leaves produce deep purple and magenta hues, and pistils change to a golden orange color. Thick trichomes grow to the tip of almost every non-fan leaf. At finish, cooler temperature and a 48-hour dark period help ripen the dense, spear-shaped buds. This hearty variety prefers temperate environments but performs well indoors or out with little maintenance. Resistance is high to pests and moderate to disease. A good choice for beginners or experienced growers, she is also suitable for commercial grows. Finishes late September or early October in Colorado at a minimum height of 4 feet (1.2m) with a max of 7 feet (2.1m).

CONSUME

The tight, compact buds of Grandma's Sugar Cookies are green and dark purple on the exterior but light green to almost white when broken open. Aromas and flavors are smooth, soft mint with complex undertones including berries, flowers, and hints of fuel. Her long-lasting euphoric effects start strong and have a slow, mellow comedown. The uplifting, happy high is clear-headed and energetic, making it good for outgoing, adventurous outdoor activities, social settings, and creative art. The stone has proven to be ill suited to unicycling, tightrope walking, or online dating. Motor vehicle and heavy machinery use should be avoided.

[photo by Ocanabis]

Killer Grape

TGA Genetics

70 S / 30 I • Giggly, Social, body relaxation
Citrus, fresh, fruity, melon

Released in early 2016, Killer Grape is the product of crossing a Killer Queen mother from British Columbia with a Querkle father (Urkle x Space Queen). Querkle, an Indica-dominant hybrid, was added to tighten spacing and lend purple traits to Killer Queen, which is a Sativa cross of the famous G-13 and Cinderella 99 strains.

GROW

Killer Grape is an indoor strain suitable for Sea of Green that prefers organic soil mix as a grow medium, so it can work outdoors as well. The plants retain much of the Killer Queen Sativa growth structure, while the Querkle Indica-ness decreases node spacing and adds a little color. This strain's phenotypes split between plants that are tall and clearly Sativa dominant and ones that are shorter, squat and more Indica-like. Queen-leaning phenos are fast growing. The breeder recommends early topping for the shorter, more Querkle-dominant phenotypes, which will need a couple extra weeks of veg time. Killer Grape's buds are generally long and spear-shaped, though some are fatter and chunkier. Her medium-sized leaves take on a blueish tinge in the Querkle-dominant phenotype. Flowering runs 8-9 weeks, with plants induced at 2 feet (0.6m) finishing at 4 feet (1.3m). Moderate to high yields can be expected with most techniques. Killer Grape has moderate resistance to adverse climactic conditions and ranks as easy to grow.

CONSUME

Killer Grape's flavors and aroma are earthy and fruity, with notes of melon and grape shining through. On the smoke, she produces an excellent-quality creeper high that is energetically upbeat and cerebral, with just enough in the body to aid relaxation but not induce sleep. Most users average 2 hours between sessions to maintain effects. The clear, focused high aids playful, creative pursuits, though almost anything goes well with Killer Grape. Medicinal users appreciate this strain's appetite stimulation and mood-elevating ability to combat depression.

Lands End
NorStar Genetics

70 S / 30 I • Alert, euphoric, trippy, sensual
• Sharp, piney, cheese

Lands End is a San Francisco spot where hippies and bikers meet to smoke a joint and enjoy the view of the Golden Gate. It's also the name of a 2014 NorStar Seeds strain that is a cross of Chem Dawg original seed stock mother from Massachusetts with a Frisco OG father bred from California's Hell's Angels OG Kush x Dead Head OG. The result nails the pungent Chem flavor while improving the structure and yield.

GROW

A versatile strain good for any type of system, Lands End retains the OG growth structure, producing an elongated pine tree shape unpruned and moderate branching when topped. These vigorous plants grow how you train them, but NorStar recommends a Low Stress Training (LST) method or a top-and-tie technique for pruning. Light organic feeding is ideal. Flowering is slow at over 9 weeks, or mid-October outdoors at temperate latitudes. Final height and size depend on pot size, but likely 4-6 feet (1.2-1.8m) at finish indoors, with moderate yields. Leaves will fade to purple and her long, rounded, dark-green buds show purple hints. Lands End is an easy-to-grow strain that's good for anyone looking for a classic Chem Dawg flavor, that's rewarding to grow for the novice, easily bouncing back from hardship when returned to good conditions.

CONSUME

The smell is strong with this one, so filters are needed if any discretion is required. Lands End flavors are complex — sharply fuel and piney, with skunky parmesan cheese, citrus, and a faint sweet curry taste. A strongly cerebral strain, Lands End clears mental confusion and leads to relaxed bliss with a bit of rubbery legs. The stone is fast acting and awake with a strong start that hits on the exhale and stays smooth with no burn out. This is a good choice for making monotonous chores fun, as well as bird watching, kayaking, massage, eating, and "couples' fun time." Not great for sleeping, naps, grocery shopping, or stock trading.

Pineapple Fields

Dynasty Genetics

80 S / 20 I • Invigorating, active • Pineapple juice and haze

The result of over a decade of work, both of Pineapple Fields' parents are Dynasty Originals that reflect generations of Professor P's careful selection. The mother is the famed Kali Snapple, a cross of Professor P's Pineapple Snowbud (ca. 2001) and an old Kali Mist male from Serious Seeds' 1990s seeds stock. The dad is a male Ms. Universe #10 dubbed the "Schwarzenegger," a cross of a refined Space Queen F3 and the ever-so-potent DesTar (Starship x Kali Mist). Ms. Universe genes shortened the Kali Snapple bloom time, making Pineapple Fields more grower-friendly.

GROW

A bushy, extremely smelly variety with thin, serrated leaves, Pineapple Fields branches so vigorously she can vine if not properly staked. Plants double in height in flowering, so stretch needs to be anticipated. This strain starts slow, but plants pack on weight after 4 weeks, stacking tight buds in spears. Some phenotypes show "foxtailing," particularly in a hot canopy, but others have well-proportioned buds that tighten nicely. Pheno "B" begins stacking after 2-3 weeks in bloom, producing swollen buds and calyxes so immense they look like seeds. Flowering time is 9-11 weeks, depending on pheno, with 25% starting slightly faster. Most buds show shades of purple and magenta with some going almost black. Drop temperatures in the last 2 weeks to deepen the hue. Pineapple Fields prefers organic nutrients, and P-K should be used sparingly in early bloom. This strain handles heat and cold well and resists most stress and problems, including powdery mildew, rot, and pests. Some beginners have had good luck, but intermediate to advanced growers will get the best outcomes and can enjoy commercial success.

CONSUME

Intense pineapple smells and flavors gave this strain her name, but discerning palates will also detect haze and funky dank along with tropical fruit. Bred for desirable Sativa effects minus the anxiety, Pineapple Fields' invigorating high imparts an instant "mental refreshment" that's productive and nearly clairvoyant. The creeper body high comes on after a few minutes and lasts 3-4 hours. This strain's well-rounded cannabinoid profile makes for effective, functional medicinal use without couch-lock. While not conducive to sleep, Pineapple Fields provides great relief for depression, anxiety, pain, and muscle spasms.

Sour Chelumbian
NorStar Genetics

80 S / 20 I • Relaxing, happy, uplifting • Sweet pine and sour citrus with earthy notes of fuel, spice, oak barrel, and sweet moss

Sour Chelumbian is an award-winning Sativa-dominant cross of Sour Diesel and Chelumbian. Her Sour Diesel mother hails from the East Coast, where she was bred from Diesel x Massachusetts Super Skunk, while her Chelumbian father is a NorStar Genetics original cross of UK Cheese x Santa Marta Columbian Gold, a California bred Sativa-dominant hybrid. NorStar selected the parents to add a sour citrus tang to the sweet, woodsy Chelumbian and improve the roots and stability of the East Coast Sour Diesel mother.

GROW

This thin-leafed strain works well indoor or outdoor, so long as her rangy, Sativa growth pattern is accommodated with proper training. The prolific branching and height is suited to a top-and-tie technique, as plants can reach over 12 feet (3.6m) unpruned outdoors, though 6 feet (1.8m) is average for indoor and greenhouse growing. Done right, this variety pumps out dense spears of long-fingered buds amidst red and yellow foliage. NorStar recommends a moderate organic fertilizer regimen. Sour Chelumbian is a great choice for both beginner and experienced gardeners. It is easy to grow in most environments, though subtle adjustments can be required for optimal yield. Finishes in 65-75 day indoors or mid-October outdoors at 38° North.

CONSUME

Sour Chelumbian exhibits a range of phenotypes from seed, from Colombian dominant to Sour Cheese, with all being winners. Her flavor is a sweet pine and sour citrus with earthy notes of fuel, spice, oak barrel, and sweet moss. The effects are quality Sativa, with an uplifting, happy high that comes on fairly fast and stays with you, making the day brighter. While not great for deep concentration or debating politics, the friendly, motivated, "go with the flow" attitude it imparts is effective as an anti-anxiety treatment.

[photo by Mono from 420Magazine]

Super Iced Grapefruit
Feminised Seeds

75 S / 25 I • Lucid, balanced, happy
• Grapefruit, lemon, citrus, berry

The parentage of Super Iced Grapefruit should come as little surprise. The Grapefruit mother, a C-99 x Sativa cross, was married to an ICE father, a Skunk x White Widow cross. Named for the ICE parent, most phenotypes of this strain also earns the name during the flowering stage with thick, frosty resin and a clear grapefruity aroma. Feminised Seeds reports that as soon as feminization techniques were reliable, the demand for feminised seeds was at least 10x higher than that of regular seeds in the European market.

GROW
Super Iced Grapefruit does best as a multi-branch plant, indoors or out, though branching is minimal and it grows mainly up, producing a central cola so large it can resemble a 2-liter soda bottle. Unpruned plants will grow to 3-4 feet (1-1.2m) indoors, with a bit of stretching in flowering, while topping creates a bushier plant amenable to training. Her dark green leaves start fat and Indica-like, but the upper leaves will thin out to show her Sativa side. Orange-haired buds swell quickly and keep expanding until ripe, when leaves start to yellow, and the resin gets so thick it brightens the green. Either organic or chemical fertilizers work well on these vigorous, temperature-tolerant plants, though use care with the watering schedule. Flowering is 8-9 weeks, or early October in northern Hemisphere. Ease of cultivation make this a good commercial strain.

CONSUME
The grapefruit aroma these plants produce in the garden carries over to a smooth citrus taste that combines lemon and grapefruit with a pungent, skunky flavor with a pleasant aftertaste that has a bit of berry. Concentrates are easy to make from the thick, heavy resin. Effects vary based on the maturity of the plants at harvest: take them early for clarity or let them go long for a more physical high. Either way, the potency (up to 22% THC) makes for a quality buzz that hits quick and sticks around. Super Iced Grapefruit is a happy, calm all-day smoke — as long as you don't have to do any serious work. Medical users report it eliminates physical pain and balances mood.

Tangie

Crockett Family Farms

70 S / 30 I • Up, active, blissful • Citrus, tangerine

Available to the public since 2014, Tangie is an aromatic cross of a Cali-O mother and a Skunk father, both from Northern California. The Cali-O mother was selected for her terpene profile and intense citrus aroma. The Skunk father was chosen for vigor and yield. The combination produces an outstanding terpene profile that has a strong tangerine aroma that saturates the air and anything that touches her.

GROW

Suitable for indoor or outdoor cultivation, this Northern California variety has been grown successfully all over the world. She prefers soil, organic fertilizers, and a multi-branch structure; producing a towering, heavily branched Christmas-tree shape that will reach 14 feet (4m) unpruned outdoors and averages 10 feet (3m). Multiple toppings of the plant are recommended. Flowering time is 9 weeks. Growth is vigorous with fast stretching that fills in during bloom. Plants that start flowering at 7 feet (2m) will finish at 12 feet (3.5m), producing an exceptional yield of long, compact spear-shaped buds frosted with trichomes and sprouting orange-hairs. The frosted lime-green buds take on a hint of purple as they ripen. Plants are mold and rot resistant and tolerate both heat and cold. Their resilience makes this a great variety for beginners and excellent for commercial propagation. Expect an intense tangerine aroma that is amazing in the garden.

CONSUME

The citrus and tangerine aroma and flavor of Tangie are so intense and penetrating that this strain qualifies as aroma therapy. Tangie has the uplifting, active, clear-headed high characteristic of Sativas. The nice and steady onset transitions smoothly to a long-lasting stone. This is not a good strain for sleep but the blissful head works well for combating depression.

[photo by Justin McIvor]

Ultra White Amnesia

Ministry of Cannabis

80 S / 20 I • Happy, active, intense • Grapefruit, candy, floral

Ultra White Amnesia adds a bit of White Widow genetics to the Amnesia Haze variety. This potent Sativa-dominant hybrid is the progeny of a marriage of Dutch and California breeding programs, crossing an Amnesia Haze mother with a White Widow x Amnesia Haze father. The resulting plants pop out acid-green calyxes amid darker green leaves, all with a thick coat of white resin.

GROW

More tall than large, this hybrid has good vigor and impressive yields. Ultra White Amnesia's mostly "up" growth pattern can be adapted to Sea of Green, but Ministry of Cannabis recommends pruning 8-10 inches (20-25cm) from bottom, leaving only four branches. Going rich in nitrogen in the first 18-20 days of flowering will enhance results. Indoors, this strain has a 9-week flowering time and benefits from having light decreased to 8 hours for the last few days. Outdoors, she appreciates a temperate, Mediterranean-type climate and finishes in the beginning of October at 42° North. The long, compact, cone-shaped buds withstand heat well and have an average sensitivity to mold. While not the best beginner strain, the challenges are modest, making this a good variety to try early or cultivate commercially.

CONSUME

The aroma of Ultra White Amnesia carries bouquets of blossoms and grape. Grapefruit and other citrus notes cut through the earthy, pungent flavors. The exceptional quality of the up, sensual high includes an intense start and long effect that reflects her THC levels of up to 22%. This happy strain complements sensual stimulation, social activities, and creativity work, so long as only low levels of attention are needed. Not great for repetitive tasks.

CBD

Cannabis contains hundreds of different chemical compounds known as cannabinoids. The two most studied and best understood are THC (Delta-9-tetrahydrocannabinol) and Cannabidiol, or CBD. THC, as most people know, is the compound responsible for the psychoactive effects (euphoria, increased creativity, increased relaxation) or "high" associated with cannabis, although it has also been shown to contain anti-inflammatory and analgesic properties. Whereas, CBD is less psychoactive, or even non-psychoactive, depending on the strain. Strains high in CBD possess significant medical benefits, including relief from inflammation, pain, anxiety, seizures, spasms and other conditions. The ratio of THC to CBD appears to play a role in determining the specific therapeutic benefits for each individual—the two work together to regulate one another and produce a synergy of effects.

Scientific and clinical research—much of which was sponsored by the US government—highlights CBD's potential as a meaningful treatment for a range of conditions, including arthritis, diabetes, alcoholism, MS, schizophrenia, depression, PTSD, antibiotic-resistant infections, and other neurological disorders. CBD has demonstrable neuroprotective and neurogenic effects, and its anti-cancer properties are currently being investigated at major academic research centers in the United States and abroad.

Many cannabis users are only now becoming aware of the benefits of CBD-rich strains and products. Several of these new strains are being crossed with old favorites leading to ever wider selection. After decades in which only high-THC cannabis was produced, numerous CBD-rich strains are now reaching cannabis users worldwide.

Ringo's Gift [photo by by Nadim Sabella]

CBD Critical Cure

Barney's Farm

80 I / 20 S • Calming, relaxing, analgesic effects • Sweet, earthiness

Barney's Farm created the Indica-dominant CBD Critical Care by crossing the legendary Indica Critical Kush with Shanti Baba's CBD Enhanced strain. Clinical studies Indicate that a balance of THC and CBD provides the most pain relief, and the cannabinoids work together to enhance other effects. This variety achieves serious levels of CBD while still packing enough synergistic THC to provide maximum therapeutic effectiveness.

GROW

This is a fast-flowering strain that produces huge, heavy buds full of resin that will need to be supported in the final weeks of flowering. The Indica characteristics come through in the plants' height, finishing at about three feet (1m). Flowering is 55-60 days, or end of September outdoors. Expect a yield of about 2 ounces per square foot (600g/m^2).

CONSUME

CBD Critical Cure has an intense and earthy flavor, with an added hint of sweetness. The CBD and THC work together beautifully to increase the therapeutic benefits. Tests on this strain Indicate you should expect about 8% CBD and 5% THC, making it an excellent choice for combatting anxiety or pain with far less psychoactive effect than most contemporary Indica strains. This variety will take the edge off what ails you.

Cannabinoid Profile

THC	CBD	CBG	THCV	CBN	CBC
4.9	7.05	0.39	0.05	0.03	0.09

[photo by Nadim Sabella]

CBD Therapy

CBD Crew

50 I / 50 S • Relaxed, happy • Earthy, truffle, fruity

CBD Therapy is a low THC and high CBD cannabis strain, solely derived from recreational high-THC strains. Fully tested in both the USA (WercShop) and Europe (Fundacion CANNA, Spain) CBD Therapy received positive feedback from growers and medical users alike. The strain took four years to stabilize through rigorous testing to ensure high CBD to THC ratios, but there may be some variability seed to seed. CBD Crew extensively tested the latest crop—they found that 50-75% of the seeds produced will have very low CBD to THC ratios (some as low as 20:1) while 25-50% of the seeds will have slightly higher THC, some as high as 2:1. Despite this variation, CBD levels will always be significantly higher than THC with all seeds.

GROW

Those looking to grow a CBD-rich plant, similar to Charlotte's Web, should consider CBD Therapy. The strain is recommended for beginning growers because it grows well both indoors and outdoors, is resistant to pests and disease and is easy to extract. Flowering time is 8-9 weeks with ripening taking place in late September or early October depending on latitude. Growing indoors may increase the THC percentage slightly. Plant height is typically 1.5 meters with a moderate to high yield depending on the conditions and the grower. The plant's shape is often wide like a pine but the shape can always be manipulated as it grows. The buds are long, bulbous and compact and many turn pink.

CONSUME

CBD Therapy's flavors range from fruity to an earthy truffle taste. There is no high to speak of, so users can perform daily activities while medicated. The effect calms nerves and reduces anxiety and worry. The unusually high CBD content in conjunction with the other cannabinoids in the plant make it perfect for those suffering from MS, Crohn's disease, fibromyalgia, inflammation, anxiety, depression or epilepsy, or those who are susceptible to the psychoactive effects of THC.

Chocolate Tonic
Purple Caper Seeds Rx

65 S / 35 I • Numbing and relaxing • Chocolate, pine, citrus

Originally a regional strain, native to the Bay Area, Chocolate Tonic is now is available worldwide. This strain originated from a CBD project designed to help patients with fibromyalgia, cancer, seizures and chronic pain. The breeders crossed a high THC father (Chocolate Kush) with a high CBD mother (Cannatonic). These strains were chosen for their high cannabinoid content, vigor and yield. The result, Chocolate Tonic, contains up to 17% CBD.

GROW

Chocolate Tonic can be grown indoors or out, in a hydro, soil or Sea of Green environment. It takes on a Christmas tree shape with little branching, so is best pruned from above. The crystals form to the tips of the leaves and frosty, compact, apple-green buds develop. Chocolate Tonic is a strong grower that can withstand heat and drought, though it is prone to powdery mildew in some environments. Because it is a hearty strain it is great for beginners and commercial growers. The strain takes 65-75 days to flower and requires 12 hours of darkness in order to do so. It has high to very high yield. Outdoor time of ripening is mid-October, in the Bay Area. Plants can reach heights of 10 feet outdoors.

CONSUME

Chocolate Tonic is a wonderful strain for pain management. Pain relief is almost instantaneous but the numbing and relaxing qualities are often sleep inducing, so users should be cautious when driving or operating heavy machinery. The medicinal effects last up to three hours. Chocolate Tonic has a chocolate, piney flavor and sometimes citrus fuel flavor.

[photo by Justin McIvor]

Ringo's Gift

Sohum Seeds

50 I / 50 S • Up, awake • Sweet, lemon

Ringo's Gift from the Southern Humboldt Seed Collective in California's Emerald Triangle boasts very high ratios of cannabidiol. This strain's mother is AC-DC, a Cannatonic phenotype from Spain, which was crossed with Harle-Tsu, a Harlequin x Sour Tsunami hybrid originating in Humboldt. The strain's name honors its breeder, Lawrence Ringo, as it was his last CBD creation and the culmination of careful breeding for consistently high ratios of CBD to THC.

GROW

This outdoor variety thrives in soil, producing long, spaced branching that benefits from regular removal of yellow leaves. She'll go 9-10 weeks of flowering with an average of 63 days, or mid October. Plants reach an average 6 feet (2m) and generate moderate yields. These thick bushes with thin light-green leaves produce chunky, compact, bulbous green and gold buds that will thank you for some supports. Ringo's Gift does well in most conditions, tolerates temperature changes, and shows some resilience and resistance to pests and diseases. All that makes this a good choice for the commercial cultivator looking for a high-CBD strain.

CONSUME

The pungent smell of Ringo's Gift in the garden becomes more sweet and lemony in the cured bud. Lemon comes through in the flavor, too, with fruity, tropical fuel dominating. The alert and awake high has a quick onset and carries the sought-after CBD properties of anxiety and pain relief. This uplifting strain is a wellness booster that is conducive to work, play, music, art, and other productive, creative activities. Research Indicates CBD acts as an antiepileptic, analgesic, anti-inflammatory, antipsychotic, antispasmodic, and neuroprotector, making this a key addition to the medicinal toolkit.

Cannabinoid Profile

THC	CBD	CBG	CBDV	CBN	CBC
0.6	9.50	0.26	0.024	0.06	0.07

Sour Tsunami 2

Sohum Seeds

50 I / 50 S • Up, alert, active
• Earthy, chocolate

A balanced strain that "washes over you like a cool wave," Sour Tsunami 2 was bred to provide pain relief without the "clogged head" or couchlock that can so often accompany the best analgesic strains. This California variety's father is a Harle-tsu (Harlequin female x Sour Tsunami male) crossed with a mother from Humboldt whose lineage includes NYC Diesel, Sour Diesel, Ferrari clone, and Albion Sour Diesel. The result is a 50/50 hybrid with serious CBD levels.

GROW

Southern Humboldt Seed Collective recommends growing Sour Tsunami 2 outdoors in the sun and soil. She will generate moderate to high yields in all climates but does better with heat than cold. The plant's dense branching forms a pine-tree shape laden with thick, long buds of green compact flowers. Any yellow leaves should be removed promptly. Expect heavy leafing 2 weeks before harvest. Flowering time is 9 weeks, with a mid-October harvest in Humboldt. Aroma in the garden is earthy but not too strong. This is a good choice for commercial cultivation.

CONSUME

Sour Tsunami 2 produces a dense flower, creating a creamy, earthy aroma and flavor with a hint of chocolate. Nine times out of ten, the plant will also produce solid CBD levels and a mild, mellow creeper high that lasts 1-2 hours. The CBD concentration can decrease inflammation, reduce anxiety, and treat many types of seizures. This is a good strain for physical activities and creative endeavors such as music and art.

Cannabinoid Profile

THC	CBD	CBG	THCV	CBN	CBC
6.0	9.82	0.39	0.30	0.03	0.21

[photo by Justin McIvor]

Suzie Q
Burning Bush Nurseries

50 I / 50 S • Therapeutic • Earthy/musky

Created in 2014 and still rare today, Suzy-Q is a CBD super star. Each of her parents were themselves bred to create a high-CBD offspring. Developed by Northern California's Project CBD from unnamed landrace genetics obtained in California but of unknown origin, Suzy-Q boasts the highest CBD level of any strain they have ever tested, with a CBD:THC ratio of 50:1. Burning Bush Nurseries named the strain Suzy Q out of a desire to "conjure up a powerful image that would evoke thoughts of marijuana and spirituality."

GROW

Suzy Q does well indoors or out but is best adapted to a mild, Mediterranean climate.

These light green plants have a mostly up, moderately bushy profile that forms a rounded cone. Suzy Q's loose, bulbous buds exhibit prominent orange hairs with touches of neon green. Leaves are thin and somewhat pointy. As a slow grower and low yielder, she does best as a multi-branch plant, showing growth patterns similar to the short internode spaces and thin branching of purple strains. The breeder recommends topping the plants once and using light nutrients with additives to emphasize her terpenes. She bounces back vigorously from pruning. Flowering time is 10 weeks, but peak CBD production is at 9 weeks. Expect plants to double in height during flowering, typically starting at 2 feet (60cm) and ending at 4 feet (1.2m), or 6-8 feet (2-2.5m) outdoors with a mild smell. This is a fine choice for commercial CBD extraction.

CONSUME

Suzy Q's flavor and aroma are earthy and musky with a touch of skunk and dark licorice. Her dominant terpene is Myrcene, the same one that gives green hops their aroma. As a CBD strain with a 50:1 CBD:THC ratio, there is no high to speak of, but Suzy-Q has an enormous capacity to alleviate a wide variety of medial aliments. This variety is particularly well-suited for treatment of cancers and seizure disorders or anyone who needs therapeutic relief but prefers to save the high for other occasions.

Strains in Alphabetical Order

[photo by Justin McIvor]

Exploring Strain Fingerprints

Founded in California in 2008, Steep Hill Labs, Inc. is a science and technology firm that has become the industry leader in cannabis testing and analytics. The company pioneered the first medical Cannabis potency and microbiological contaminants testing methodology for use in California—the first state to legalize medical cannabis.

Steep Hill currently has compiled data from over one hundred thousand flower samples to produce the industry recognized strain fingerprints. Steep Hill currently tests for 17 cannabinoids and up to 43 terpenes, but the data reported here were collected prior to the release of the 43 terpene test, and so only reflect the 10 terpenes measured by our LC method. Strain Fingerprints are the statistical analysis of the combined data from tests for the same named strains, where the number of independent tests are 10 or greater. The data are reported as the range of the average percent by mass of a particular analyte (e.g., cannabinoid or terpene), typically reported as per gram of flower. For this publication, some strain fingerprints were produced without the required number of independent tests (Ringo's Gift). For the purposes of this publication, the Ringo's Gift strain fingerprint was created using only five independent samples, instead of the minimum number of 10.

In addition to reporting the data to the client, the data are used to investigate differences in not only final levels of cannabinoids and terpenes, but also the ratios of the cannabinoids and terpenes to each other. This information can be used to gain a better understanding of the differences between strains, or at least between lineages, Steep Hill's quest to find out what really makes one strain different from another.

Steep Hill urges everyone to embrace testing. The more data we have on the cannabinoid and terpene content of cannabis, the better armed we are to use the plant both medicinally and recreationally. Only through testing can we build better strains, identify what strains help what conditions and figure out what makes one strain or lineage different from the others. Testing helps identify which strains, when bred together, can produce better strains or allows enhancement of specific traits. With that level of understanding, we can truly develop cannabis as a medicine worthy of the respect held for other medicines. And while much of what Steep Hill does is to make cannabis safer, and better characterized, we aren't strangers to the benefits of more and better testing to the recreational aspects of cannabis as well. No matter how you look at it, testing helps make better cannabis.

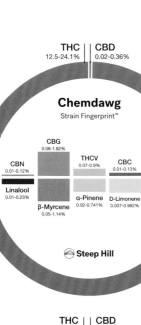

Chemdawg
Strain Fingerprint™

THC
12.5-24.1%

CBD
0.02-0.36%

CBG
0.08-1.82%

CBN
0.01-0.12%

THCV
0.07-0.9%

CBC
0.01-0.13%

CBL
0.16-0.67%

Linalool
0.01-0.23%

β-Myrcene
0.05-1.14%

α-Pinene
0.02-0.741%

D-Limonene
0.037-0.982%

β-Caryophyllene
0.02-1.4%

Steep Hill

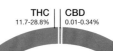
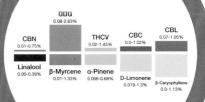

Girl Scout Cookies
Strain Fingerprint™

THC
11.7-28.8%

CBD
0.01-0.34%

CBG
0.08-2.63%

CBN
0.01-0.75%

THCV
0.02-1.45%

CBC
0.0-1.02%

CBL
0.07-1.05%

Linalool
0.00-0.39%

β-Myrcene
0.01-1.33%

α-Pinene
0.006-0.68%

D-Limonene
0.019-1.3%

β-Caryophyllene
0.0-1.13%

Steep Hill

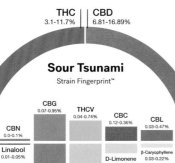

Sour Tsunami
Strain Fingerprint™

THC
3.1-11.7%

CBD
6.81-16.89%

CBG
0.07-0.95%

CBN
0.0-0.1%

THCV
0.04-0.74%

CBC
0.12-0.36%

CBL
0.03-0.47%

Linalool
0.01-0.05%

β-Myrcene
0.06-1.46%

α-Pinene
0.141-1.008%

D-Limonene
0.088-0.531%

β-Caryophyllene
0.03-0.22%

Steep Hill

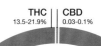
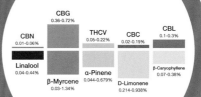

True OG
Strain Fingerprint™

THC
13.5-21.9%

CBD
0.03-0.1%

CBG
0.36-0.72%

CBN
0.01-0.06%

THCV
0.05-0.22%

CBC
0.02-0.19%

CBL
0.1-0.3%

Linalool
0.04-0.44%

β-Myrcene
0.03-1.34%

α-Pinene
0.044-0.679%

D-Limonene
0.214-0.938%

β-Caryophyllene
0.07-0.38%

Steep Hill

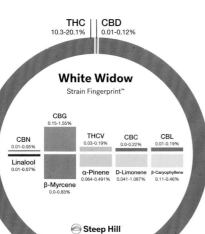

White Widow
Strain Fingerprint™

THC
10.3-20.1%

CBD
0.01-0.12%

CBG
0.15-1.55%

CBN
0.01-0.05%

THCV
0.03-0.19%

CBC
0.0-0.22%

CBL
0.01-0.19%

Linalool
0.01-0.07%

β-Myrcene
0.0-0.83%

α-Pinene
0.064-0.491%

D-Limonene
0.041-1.087%

β-Caryophyllene
0.11-0.46%

Steep Hill

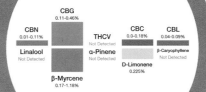

Ringo's Gift
Strain Fingerprint™

THC
0.4-0.8%

CBD
5.55-11.63%

CBG
0.11-0.46%

CBN
0.01-0.11%

THCV
Not Detected

CBC
0.0-0.18%

CBL
0.04-0.09%

Linalool
Not Detected

β-Myrcene
0.17-1.18%

α-Pinene
Not Detected

D-Limonene
0.225%

β-Caryophyllene
Not Detected

Steep Hill

[photo by Gracie Malley]

Sponsors

Thanks to all of the businesses, organizations and individuals that supported this project.

Enhancing your life through cannabis.

Hydrolife focuses on connecting growers, medical practitioners, patients, and health enthusiasts.

Distributed to growers, hydro shops, dispensaries, medical clinics and more, Hydrolife connects
the interests and passions of the industry with cutting-edge medical advancements.

grow. heal. live. enjoy.

myhydrolife.com

Trusted for over 25 years
Providing Quality Nutrient Products for Soil and Hydroponics

Elements

SeaBlast™

Bloom Master™

BioRighteous™

Sugar Peak™

BioZeus™

GodSilica™

Sweet & Heavy™

OilyCann

Charts and more details, visit **earthjuice.com**

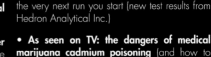

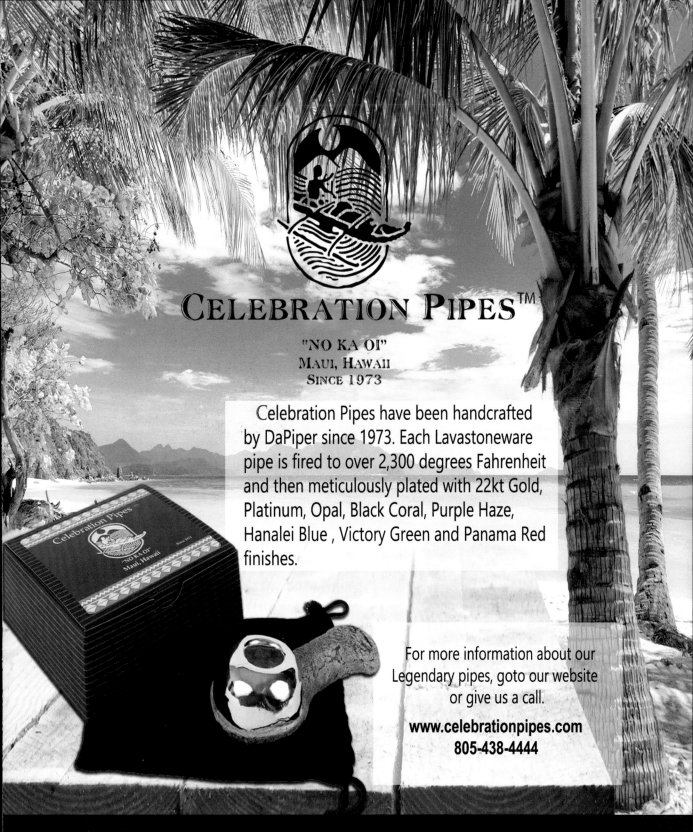

ED Rosenthal
Grow with the best

Photo by Dabsel Adams

www.edrosenthal.com